AUTOCAD 2012
BEGINNING
AND
INTERMEDIATE

AUTOCAD 2012
BEGINNING
AND
INTERMEDIATE

By
Munir M. Hamad
Autodesk™ Approved Instructor

MERCURY LEARNING AND INFORMATION
Dulles, Virginia
Boston, Massachusetts

Publisher: David Pallai

MERCURY LEARNING AND INFORMATION LLC
22841 Quicksilver Drive
Dulles, VA 20166
info@merclearning.com
www.merclearning.com
1-800-758-3756

This book is printed on acid-free paper.

Munir M. Hamad, AutoCAD™ 2012 Beginning and Intermediate.
ISBN: 978-1-936420-20-9

The publisher recognizes and respects all marks used by companies, manufacturers, and developers as a means to distinguish their products. All brand names and product names mentioned in this book are trademarks or service marks of their respective companies. Any omission or misuse (of any kind) of service marks or trademarks, etc. is not an attempt to infringe on the property of others.

Library of Congress Control Number: 2011931631

111213321

Printed in Canada

Our titles are available for adoption, license, or bulk purchase by institutions, corporations, etc. For additional information, please contact the Customer Service Dept. at 1-800-758-3756 (toll free).

CONTENTS

Chapter 2: Precise Drafting in AutoCAD 2012

Chapter 3: Modifying Commands Part I

Chapter 5: Layers and Inquiry Commands

Chapter 6: Blocks and Hatches

Chapter 7: Text and Tables

Chapter 8: Dimensions

Chapter 9: Plotting

ABOUT THE BOOK

This book is for new and novice users of AutoCAD 2012. It covers the basic and intermediate levels. The book takes the reader on a journey to mastering AutoCAD techniques using nine chapters that cover everything you need to create a full engineering drawing starting from preparing the drawing and ending with how to plot it. The tenth chapter covers three full projects (metric and imperial) for architectural (one project) and mechanical (two projects) designs.

A summary of each chapter follows:

- Chapter 1 covers the basics of AutoCAD along with the interface.
- Chapter 2 covers AutoCAD techniques for drawing with accuracy.
- Chapters 3 & 4 cover all the modifying commands and show the user how to modify and construct a drawing.
- Chapter 5 covers the AutoCAD method of organizing a drawing using layers and inquiry commands.
- Chapter 6 covers the methods of creating and editing blocks and inserting and editing hatches.
- Chapter 7 covers AutoCAD methods of writing text and inserting tables.
- Chapter 8 covers how to create and edit dimensions in AutoCAD.
- Chapter 9 covers how to plot a drawing.
- Chapter 10 includes three projects: one architectural design and two mechanical designs covering both metric and imperial units.

PREFACE

- Since its inception, AutoCAD has enjoyed a wide user base and has become the most-used CAD software since the 1980s. This widespread use is due to its simplicity, which makes it very easy to learn.
- AutoCAD has evolved through the years and has become very comprehensive, addressing all aspects of engineering and architectural drafting and designing.
- This book addresses basic and intermediate levels of AutoCAD techniques, which makes it ideal for novice AutoCAD users and students in colleges and universities.
- This book is not a replacement for the manual(s) that comes with the software but is complementary to it with its 57 practices for strengthening and solidifying the techniques and skills discussed.
- Solving all practices is essential for the user because, in the end, AutoCAD is practical and not theoretical; it has to be learned.
- At the end of each chapter the reader will find "Chapter Review Questions," which include questions you may see in an Autodesk exam. Odd answers are provided at the end of each chapter.
- Chapter 10 contains three projects, and one is an architectural plan. This project will allow users to master the knowledge needed to land a job in today's job market. The other two projects are for mechanical engineering. One of them is thoroughly explained and the other is not! These projects as well as the others are presented in both metric and imperial units.

ABOUT THE DVD

- The DVD included with this book contains:
    - A link to the AutoCAD 2012 trial version, which will last for 30 days starting from the day of installation. This version will help you solve all practices and projects.
    - Practice files, which will be your starting point for solving all the practices in the book. The "Practices" folder should be copied onto one of the hard drives of your computer.
    - You will also find two "Projects" folders: one is called "Metric" for metric unit projects, and the second one is called "Imperial" for imperial unit projects.

PREREQUISITES

- The author assumes that the reader has experience using computers and has working experience using the Windows Operating System. The reader should know how to start a new file, open an existing file, save, and use save as as well as close without saving and exiting the software. These commands are almost the same in all software packages, including AutoCAD.
- AutoCAD 2012 has a dark gray background, but the screenshots in this book have been changed to white for better visibility.

Chapter 1

AutoCAD 2012 Basics

In This Chapter

◇ How to start AutoCAD
◇ How to work with the AutoCAD interface
◇ What are the AutoCAD defaults and drawing units?
◇ How to work with file-oriented commands
◇ Undo and Redo commands

1.1 HOW TO START AUTOCAD

- AutoCAD was released in 1982 by Autodesk, Inc. (a small company at the time) and designed for PCs only. Since then AutoCAD has boasted the biggest user base in the world in the CAD industry. You can use AutoCAD for both 2D and 3D drafting and designing and for architectural, structural, mechanical, electrical, road and highway designs, and environmental and manufacturing drawings.
- Though the theme these days is BIM (Building Information Modeling) AutoCAD is still the most profitable software for Autodesk, Inc. due to its ease-of-use and totality.
- There is another version of AutoCAD, called AutoCAD LT, used only 2D drafting.
- To start AutoCAD 2012, double-click the shortcut on your Desktop (created in the installation process), and AutoCAD will show the **Autodesk Exchange** window, which looks like the following:

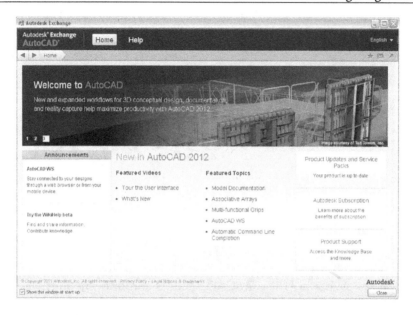

- Under **Featured Videos**, take a look at the number of available videos; these videos will change each day. You can also view discussions on the **Featured Topics**. At the right, you will see links to **Product Updates and Service Packs**, **Autodesk Subscription**, and **Product Support**. You can also select not to **Show this window at start up**. When done, click the Close button.
- The AutoCAD screen contains the following parts:

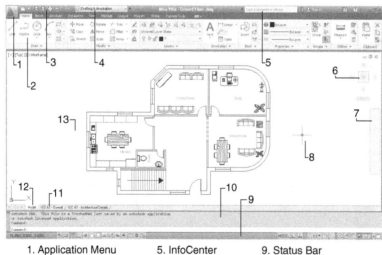

1. Application Menu	5. InfoCenter	9. Status Bar
2. Ribbon	6. ViewCube	10. Command Window
3. Quick Access Toolbar	7. Navigation Bar	11. Layout
4. Workspace	8. Crosshairs	12. Model

1.2 AUTOCAD 2012 INTERFACE

- This interface is similar to the Microsoft Office 2007/2010 interface. You will mainly use the Ribbons and Application Menu to reach commands. The most important feature of this new interface is the size of the **Graphical Area**, which is much larger.

1.2.1 Application Menu

- The Application Menu contains the file-related commands:

- Commands such as creating a new file, opening an existing file, saving the current file, saving as the current file under a new name and different folder, exporting the current file to a different file format, printing and publishing the current file, etc., are found here. Most of these commands will be discussed throughout the book.

1.2.2 Quick Access Toolbar

- This toolbar contains all the File commands in the Application Menu along with Workspace and Undo/Redo.

- You can customize this toolbar by clicking the arrow at the right. You will see the following:

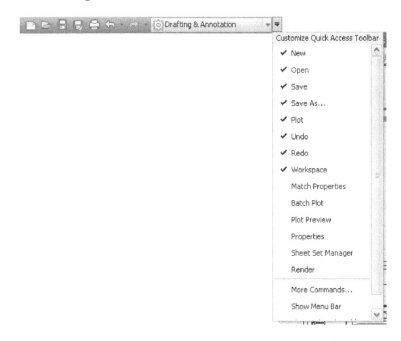

- As you can see, you can add or remove commands and choose **Show Menu Bar**, which you may find useful sometimes.

1.2.3 Ribbons

- Ribbons consist of two parts: tabs and panels, as shown below:

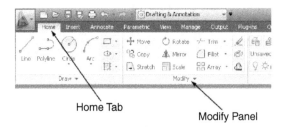

Home Tab

Modify Panel

- Some panels have more buttons than shown. The following is the **Modify** panel:

- Click the small triangle near the title, and you will see the following:

- If you move away from the panel, the buttons will disappear. To make them visible again click the push pin, and the new view will look like the following:

- For some commands, there are many options. To make things easier, AutoCAD put all the options with the corresponding button. See the following illustration:

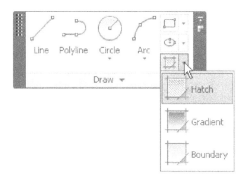

- Ribbons have a very simple but useful help feature. If you move the mouse over a button a small help screen will pop up:

- If you leave the mouse pointer over the button, AutoCAD will show more detailed help:

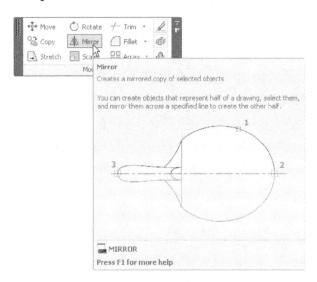

- Other buttons have video help, like the following:

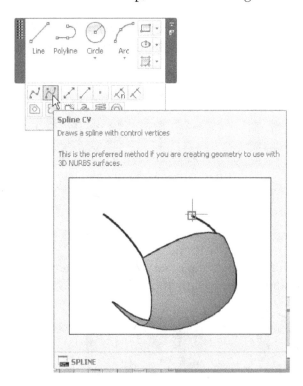

- Panels have two states, *docked or floating*. By default, all panels are docked in their respective tab. Drag and drop the panel in the graphical area to make it floating. If a panel is floating, you will be able to see it while other tabs are active.
- You can send the panel back to its respective tab by clicking the small button at the top-right side:

- While the panel is floating you can toggle the orientation:

- It will either extend to the right:

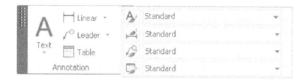

- Or down:

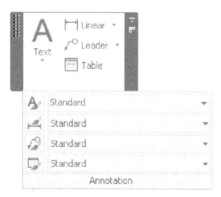

- The small arrows at the end of the tabs allow you to cycle through the different states of the Ribbons. The main objective of this new feature is to give you yet more graphical area to work in. Clicking the small arrow will show the following:

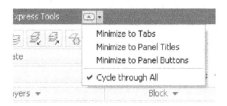

1.2.4 InfoCenter

- The **InfoCenter** is the place to find help topics online and offline, along with other helpful tools:

- For example, if you type a word or phrase in the field shown, AutoCAD will open the Autodesk Exchange window and find all the related topics online and offline. Online means it will search all Autodesk websites along with some popular blogs.
- Sign In will allow you to sign into Autodesk Online Services. The X at the right will activate the Autodesk Exchange window and show you the latest information and video tutorials. The last button at the right with the question mark will show the following:

1.2.5 Command Window

- Reading the Command window will help you understand what AutoCAD wants from you. AutoCAD will show two things in the Command window: your commands and AutoCAD prompts asking you to do something such as specify a point, an angle, etc. See the following illustration:

```
Command: LINE
Specify first point:
Specify next point or [Undo]:
Command:
```

1.2.6 Graphical Area

- The graphical area is your drafting area. This is where you will draw all your lines, arcs, circles, etc. It is a precise environment with X, Y, Z space for 3D and the X, Y plane for 2D. You can monitor coordinates in the left part of the status bar.

1.2.7 Status Bar

- The status bar in AutoCAD contains coordinates along with important functions; some of the functions are for precise drafting and some of them are for other more advanced features such as Infer Parameters, Cycling, and Transparency.

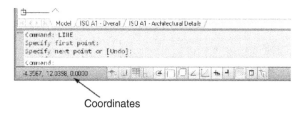

Coordinates

1.3 AUTOCAD DEFAULTS

- There are some settings you should be familiar with before working in the AutoCAD environment. These include:
  - AutoCAD saves points as Cartesian coordinates (X,Y) for both metric and imperial numbers. This is the first method of precise input in AutoCAD, typing the coordinates using the keyboard:

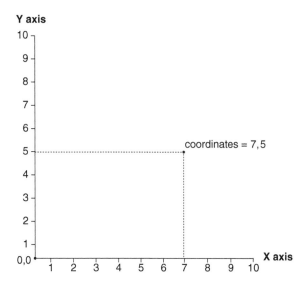

- To specify angles in AutoCAD, assume East (to your right) is 0° and then go counter-clockwise. (This is applicable only for the northern hemisphere, but we will learn in Chapter 3 how can we change this setting for the southern hemisphere.)

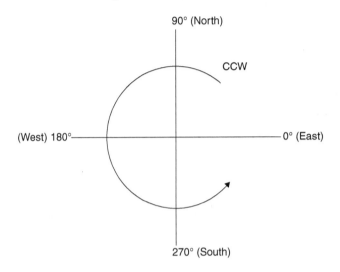

- The mouse wheel has four zooming functions: Zoom In (moving wheel forward), Zoom Out (moving wheel backward), Panning (pressing and holding the wheel), and Zooming Extents (double-clicking the wheel).
- Pressing [Enter] or [Spacebar] is the same in AutoCAD.
- Pressing [Enter] without typing any command in AutoCAD will repeat the last command. If it is the first thing you do in the current session, it will start Help.
- Pressing [Esc] will cancel the current command.
- Pressing [F2] will show the Text Window.

1.4 WHAT IS MY DRAWING UNIT?

- If you draw a 6-unit line in AutoCAD, what units will AutoCAD use? Will it be 6 m or 6 ft. or neither? AutoCAD works with all types of units; if you want to use 6 m. AutoCAD will do this. If you mean 6 ft AutoCAD will work with this unit as well. You just need to be consistent with your measurements throughout the file. While this is true in Model Space where you will do your drafting, when it comes to printing, you will have to make sure to set your drawing scale correctly. In Chapter 9, we will discuss printing.

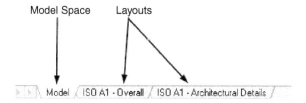

1.5 CREATING A NEW AUTOCAD DRAWING

- This command will allow you to create a new drawing based on a pre-made template (we will discuss how to create your own template in Appendix A). Use the **Quick Access Toolbar** and click the **New** button:

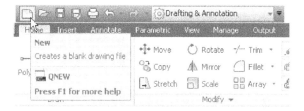

- You will see the following dialog box:

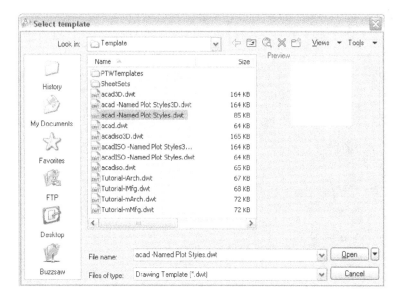

- Take the following steps:
    - Select the desired template file (AutoCAD template files have a *.dwt* extension). AutoCAD 2012 comes with many pre-made templates you can use (it is actually better, however, to create your own template files).
    - Once you are done, click the **Open** button.
    - AutoCAD drawing files have a *.dwg* file extension.
    - AutoCAD will start with a new file with a temporary name such as *Drawing1.dwg*, but you should rename it to something meaningful.

1.6 OPENING AN EXISTING AUTOCAD DRAWING

- This command will allow you to open an existing drawing file for additional modifications. From the **Quick Access** Toolbar, click the **Open** button:

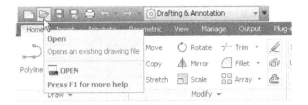

- You will see the following dialog box:

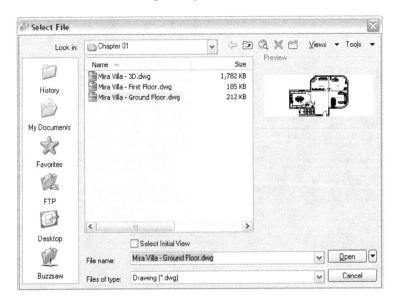

- Take the following steps:
  - Specify your desired drive and folder.
  - You can open a single file by selecting its name from the list and clicking the **Open** button, or you can double-click on the file's name. Or, you can open more than one file by selecting the first file name, then holding the [Ctrl] key and clicking the other file names in the list (a common Windows shortcut) and then clicking the **Open** button.

1.6.1 Quick View

- Using the status bar, and while multiple files are open, you can use one of two functions: **Quick View Drawings** and **Quick View Layouts**:

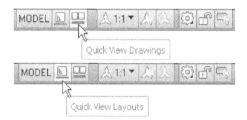

- **Quick View Drawings** will show you something like the following:

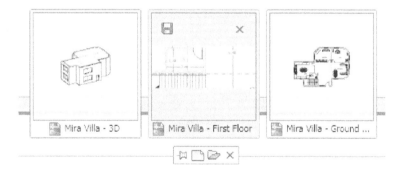

- These are small viewing windows for all of your open files. Clicking any of the small views will allow you to jump to the file and see the layouts in it. See the following:

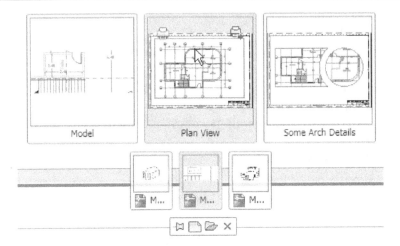

- A small toolbar at the bottom will appear as well, which will help you to:
  - Close the **Quick View** window
  - Open another file
  - Start a new drawing
  - Use the push pin to make **Quick View Drawings** permanent
- Right-clicking in the **Quick View** window will allow you to see the following menu (which is self-explanatory):

1.6.2 Organizing Files

- If you have more than one file open, you can arrange them on the screen. Using the **View** tab and **Window** panel you can use several methods to organize your files including tiling horizontally, vertically, and cascading:

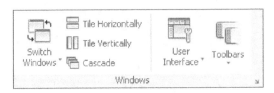

- Alternatively, the Switch Drawings button will show you a list of all opened files, and you can click the file you need. Another way to switch between files is using [Ctrl] + [Tab].

1.7 CLOSING DRAWING FILE(S)

- This command will allow you to close the current opened file(s), or all opened files depending on the command you choose. Use the **Application Menu** and move your mouse to the **Close** button, then select either **Current Drawing** to close the current file or **All Drawings** to close all the opened files:

- If any of the opened files was modified, AutoCAD will ask if you want to save or close without saving. See the following dialog box:

1.8 UNDO AND REDO COMMANDS

- The Undo and Redo commands help you correct mistakes. They can be used in the current session only.

1.8.1 Undo Command

- This command will undo the effects of the last command. You can reach this command by going to the **Quick Access** toolbar and clicking the **Undo** button. If you want to undo several commands, click the small arrow at the right. You will see a list of the commands; select the group and undo them:

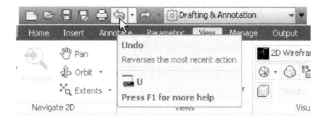

- You can also type **u** at the Command window (don't type undo because it has a different meaning) here, or press [Ctrl] + Z on the keyboard.

1.8.2 Redo Command

- This command will undo the Undo command. You can reach this command from the **Quick Access** toolbar by clicking the **Redo** button. If you want to redo several commands, click the small arrow at the right. You will see a list of recent commands; select the group and redo them:

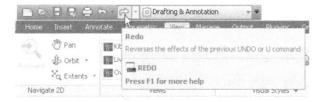

- You can also type **redo** at the Command window, or press [Ctrl] + Y on the keyboard.

NOTES:

AUTOCAD BASICS

Practice 1-1

1. Start AutoCAD 2012.
2. Open the following files:
   a. **Mira Villa – Ground Floor**
   b. **Mira Villa – First Floor**
   c. **Mira Villa – 3D**
3. Using the **Quick View Layouts** on the status bar, take a look at the three files and their layouts.
4. Using [Ctrl] + [Tab] browse the three files.
5. Use the different zoom techniques with the mouse wheel.
6. Tile the three files vertically.
7. Using the **Application Menu**, close all files without saving.

CHAPTER REVIEW

1. **Quick View** allows you to see the Model Space and layouts of open files.
   a. True
   b. False
2. AutoCAD template files have the _____ extension.
   a. °.dwt
   b. °.dwg
   c. °.tmp
   d. °.temp
3. AutoCAD units can be meters or feet, whichever you choose.
   a. True
   b. False
4. Moving the mouse wheel forward will _____.
5. To undo any command in AutoCAD you can:
   a. Click the Undo icon from the **Quick View** toolbar
   b. Type **u** in the Command window
   c. Type [Ctrl] + Z
   d. All of the above
6. Ribbons consist of _____ and _____.

7. The menu bar is not shown by default, but you can make it visible.
   a. True
   b. False
8. The AutoCAD drawing file extension is _____.
9. Positive angles in AutoCAD are _____.

CHAPTER REVIEW ANSWERS

1. a
3. a
5. d
7. a
9. CCW

PRECISE DRAFTING IN AUTOCAD 2012

Chapter **2**

In This Chapter
◇ Drafting priorities
◇ How to draw lines, circles, and arcs using precise methods
◇ How to draw polylines using precise methods
◇ How to convert lines and arcs to polylines and vice-versa
◇ Object Snap and Object Track

2.1 DRAFTING PRIORITIES

- There are two main drafting priorities for most people: accuracy and speed. Most people want to finish their drawings fast, but without compromising accuracy. Experts tend to put accuracy first, but at the expense of speed.
- In this chapter you will learn how to use the four most important drafting commands. These are:
  - The Line command, used to draw line segments.
  - The Arc command, used to draw circular arcs.
  - The Circle command, used to draw circles.
  - The Polyline command, used to draw lines and arcs jointly.
- While we are discussing the four drafting commands, we will also cover accuracy tools and tools to help you speed up the drafting process.

2.2 DRAWING LINES USING THE LINE COMMAND

- This command will enable you to draw straight lines; each line segment presents a single object. To issue this command go to the **Home** tab, locate the **Draw** panel, then select the **Line** button:

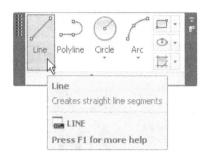

- You will see the following AutoCAD prompts:

```
Specify first point:
Specify next point or [Undo]:
Specify next point or [Undo]:
Specify next point or [Close/Undo]:
```

- Using the first prompt specify the coordinates of your first point. Continue specifying points until you are done, and keep the following in mind:
  - If you want to stop without closing the shape, simply press [Enter]. ([Esc] will do the job as well, but don't make it a habit, since [Esc] generally means abort.)
  - If you want to close the shape and finish the command, press **C** on the keyboard, or right-click and select the **Close** option.
  - If you make a mistake you can undo the last point by typing **u** on the keyboard, or right-clicking and selecting the **Undo** option.
- This is what the right-click menu looks like:

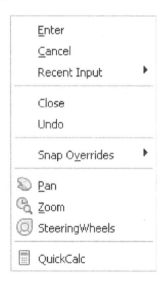

2.3 WHAT IS DYNAMIC INPUT IN AUTOCAD?

- Dynamic Input has multiple functions:
  - It will show all the prompts in the Command window in the graphical area.
  - It will show the lengths and angles of the lines before drafting, which will allow you to specify them accurately.
- In order to turn on/off **Dynamic Input** use the following button on the status bar:

2.3.1 Example for Showing Prompts

- By default, if you type any command using the Command window, Auto-CAD will help by showing all the commands starting with the same word(s). See the following example:

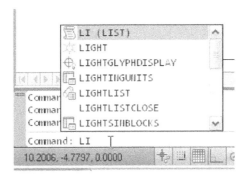

- We typed the two letters **LI** and accordingly AutoCAD gave us all the commands starting with these two letters. While Dynamic Input is on, this is applicable to the crosshairs as well. You will see the following:

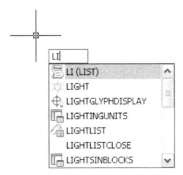

- Select the Line command from the list. Once you press [Enter], the following prompt will appear:

- Type in the X and Y coordinates, using the [Tab] key to move between the two fields:

2.3.2 Example for Specifying Lengths and Angles

- Once you have specified the starting point, AutoCAD will use Dynamic Input to show the length and the angle of the line using **Rubberband** mode:

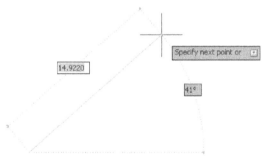

- Angles are measured CCW starting from the East, but only for 180°, unlike the angle system in AutoCAD, which is for the whole 360°.
- Type the length of the line, then using [Tab] input the angle (it will increase in 1° increments). Once you are finished, press [Enter] to specify the first line; continue doing the same for the other segments:

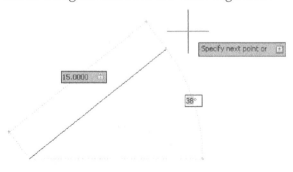

DRAWING LINES USING DYNAMIC INPUT

 Practice 2-1

1. Start AutoCAD 2012.
2. Open the file **Practice 2-1.dwg.**
3. Using the status bar, click off Polar Tracking, Ortho, Object Snap, and make sure Dynamic Input is on.
4. Draw the following shape, using 0, 0 as your starting point, keeping in mind that all sides = 4 and all angles are multiples of 45°:

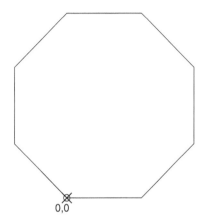

5. Save the file and close it.

2.4 EXACT ANGLES (ORTHO VS. POLAR TRACKING)

- Dynamic Input angles are incremented by 1°, but you cannot depend on it to specify angles precisely in AutoCAD.
- The **Ortho** function will force the lines to be at right angles (orthogonal) using the following angles: 0, 90, 180, and 270.
- In order to turn on/off **Ortho** use the following button on the status bar:

- But what if you want to use other angles such as 30, 45, 60, etc.? For this reason AutoCAD introduced another function called Polar Tracking, which allows you to see in the graphical area <u>rays</u> starting from the current point pointing towards angles such as 30, 45, etc., and based on the settings, you can specify angles. Since Ortho and Polar Tracking contradict each other, when you switch to one, the other will be turned off automatically.

- To turn on/off **Polar Tracking** use the following button on the status bar:

- When you right-click the button on the status bar, you will see the following menu:

- You can select the desired angle or select **Settings** to change some of the default settings of Polar Tracking. You will see the following dialog box:

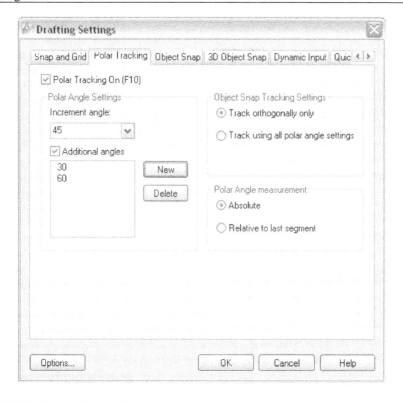

2.4.1 Increment Angle

- The Increment angle is the angle to be used along with its multiples. Select one from the list or type your own.

2.4.2 Additional Angles

- If you are using 30 as your increment angle, then 45 will not be among the angles that Polar Tracking will allow you to use. So, you will need to specify it as an additional angle. But be aware that you will not use its multiples.

2.4.3 Polar Angle Measurement

- When you are using Polar Tracking, you have the ability to specify angles as absolute angles (based on 0° at the East) or use the last line segment as your 0 angle. See the following illustration:

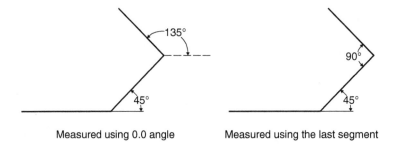

Measured using 0.0 angle   Measured using the last segment

 NOTE ■ While you control the angle using either Ortho or Polar Tracking, you can type in the distance desired, then press [Enter]. This will allow you to draw accurate distances. This method is called *Direct Distance Entry*.

EXACT ANGLES

Practice 2-2

1. Start AutoCAD 2012.
2. Open **Practice 2-2.dwg**.
3. Draw the following shape (without dimensions) using the Line command and 0,0 as your starting point, keeping in mind that you have to use Polar Tracking. Set the proper Increment angle and additional angles using the Direct Distance Entry method to input the exact distances:

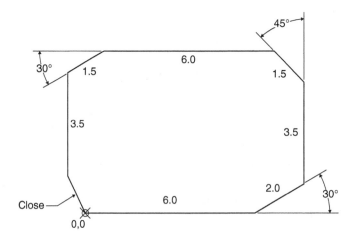

4.   Save and close the file.

2.5 USING SNAP AND GRID TO SPECIFY POINTS ACCURATELY

- Snap and Grid is another method of specifying points accurately in the X,Y plane.
- Using the mouse alone (the default method) is not accurate, so we can't depend on it to specify points. We need to control its movement, which is the sole function of Snap. Snap can control the mouse and make it jump in the X and Y directions in exact distances.
- Grid by itself is *not* an accurate tool, but it does complement the Snap function. It will show horizontal and vertical lines replicating the drawing sheets. In order to turn on/off the **Snap** tool use the following button on the status bar:

- In order to turn on/off the **Grid** tool use the following button on the status bar:

- Most of the time, using both tools at the same time will not be enough, but you can modify the settings to fit your needs. Right-click one of the two buttons and the following shortcut menu will appear:

- Select the **Settings** option and the Drafting Settings dialog box will pop up:

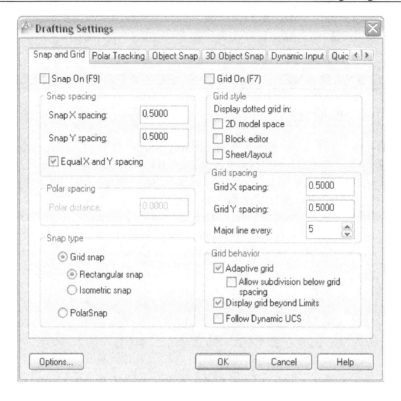

- Input the Snap X Spacing and Snap Y Spacing (by default they are equal). Switch off the checkbox to make them unequal. Do the same for the Grid Spacing in X and Y. If you want Grid to follow Snap, set the Grid spacing to zeros. In Grid there are major and minor lines; set the major line frequency.
- Set if you want to see the Grid in dots and where (2D model space, Block editor, or Sheet layout).
- Grid behavior is for 3D only.
- Specify the type of Snap: Grid Snap or Polar Snap. Polar Snap will work with Polar Tracking and allow you to specify exact distances in angles such as 30, 45, 60, etc. You can also use the following function keys to turn on/off both Snap and Grid:
  - F9 = Snap on/off
  - F7 = Grid on/off

SNAP AND GRID

 Practice 2-3

1. Start AutoCAD 2012.
2. Open **Practice 2-3.dwg**.
3. Switch both Snap and Grid on.
4. Set the Snap to be 1 in both X and Y, and set Grid to follow it.
5. Draw the following shape:

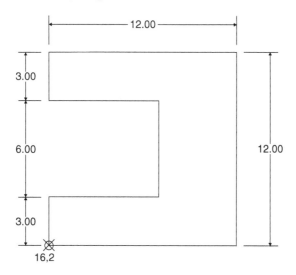

6. Save and close the file.

2.6 DRAWING CIRCLES USING THE CIRCLE COMMAND

- This command will draw a circle using different methods based on the available data. If you know the coordinates of the center, there are two available methods. If you know the coordinates of points at the parameter of the circle, there are two additional methods. Finally, if there are drawn objects such as lines, arcs, or other circles that can be used as tangents for the to-be-created circles, there are two more methods. There are six methods for drawing a circle in AutoCAD:

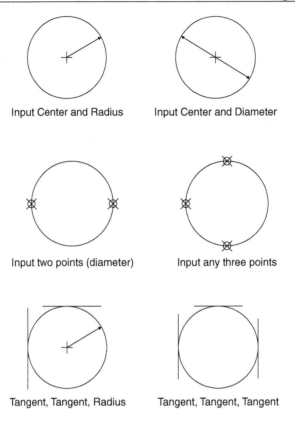

Input Center and Radius   Input Center and Diameter

Input two points (diameter)   Input any three points

Tangent, Tangent, Radius   Tangent, Tangent, Tangent

- To issue this command, go to the **Home** tab, locate the **Draw** panel, then select the arrow near the **Circle** button to see all the available methods:

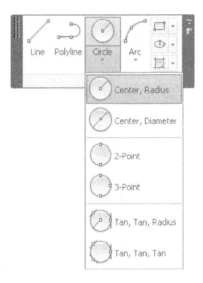

2.7 DRAWING CIRCULAR ARCS USING THE ARC COMMAND

- ▪ This command will draw an arc part of a circle. To make our lives easier, AutoCAD allows us to use eight pieces of information related to a circular arc. These are:
  - • The starting point of the arc.
  - • Any point as a second point on the parameter of the arc.
  - • The ending point of the arc.
  - • The direction of the arc, which is the tangent that passes through the start point. You should input the angle of the tangent.
  - • The distance between the starting point and the ending point, which is called the Length of Chord.
  - • The center point of the arc.
  - • The radius.
  - • The angle between Start-Center-End, which is called the Included Angle.
- ▪ See the following illustration:

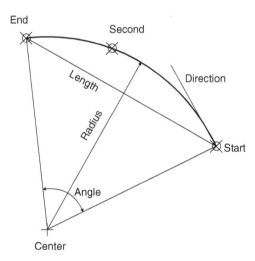

- ▪ If you provide three of these eight, AutoCAD will be able to draw an arc, but you cannot provide just any three. The combination of the information needed can be found in the **Home** tab, using the **Draw** panel, while clicking the arrow near the **Arc** button to see all the available methods:

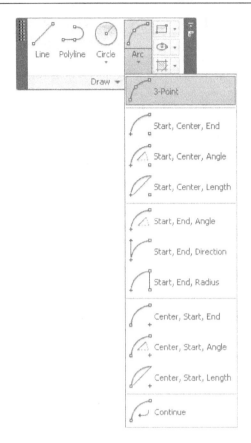

- As you can see, the Start point is always one of the required pieces of information. You should think CCW when specifying points.

DRAWING CIRCLES AND ARCS

Practice 2-4

1. Start AutoCAD 2012.
2. Open **Practice 2-4.dwg**.
3. Switch on Snap only, and make sure Object Snap is off.
4. Draw the four arcs as shown below using any of the methods you learned.
5. At point 27,12 draw a circle with radius = 1.
6. At point 17,12 draw another circle with the same radius.
7. Using the Circle command and the 2-Points method, specify a circle between the two points 25,7 and 23,7.

8. Do the same thing again, but this time using 21,7 and 19,7.
9. Switch off Snap.
10. Using the Circle command and Tan, Tan, Radius, use the two circles you drew in steps 7 and 8 as your tangents (click the circle at the right from its right and the circle at the left to its left), then specify radius = 3.
11. You should have the following shape:

12. Save and close the file.

2.8 DRAWING LINES AND ARCS USING THE POLYLINE COMMAND

- The Polyline command will allow you to do all or any of the following:
  - Draw both line segments and arc segments.
  - Draw a single object with the same command rather than drawing segments of lines and arcs, as with the Line and Arc commands.
  - Draw lines and arcs with starting and ending widths.
- To use the command go to the **Home** tab, locate the **Draw** panel, then select the **Polyline** button:

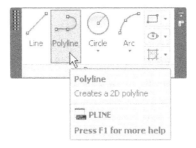

- The following prompt will appear:

```
Specify start point:
Current line-width is 1.0000
Specify next point or [Arc/Halfwidth/Length/Undo/
Width]:
```

- AutoCAD will ask you to specify the first point, and when you do, AutoCAD will report to you the current line-width. If you like it, continue specifying points using the same method we learned with the Line command; if not, change the width as a first step by typing the letter **w**, or right-clicking and selecting the **Width** option. You will see the following prompt:

```
Specify starting width <1.0000>:
Specify ending width <1.0000>:
```

- Specify the starting width, press [Enter], and then specify the ending width. Next time you use the same file, AutoCAD will report these values to you when you issue the Polyline command. Halfwidth is the same, but instead of specifying the full width, you specify the halfwidth.
- The Undo and Close options are identical to the ones in the Line command.
- Length will allow you to specify the length of the line using the angle of the last segment.
- Arc will allow you to draw an arc attached to the line segment. You will see the following prompt:

```
Specify endpoint of arc or [Angle/Center/Close/
Direction/ Halfwidth/Line/Radius/Second
pt/Undo/Width]:
```

- The arc will be attached to the last segment of the line or will be the first object in a Polyline command. Using either method, the first point of the arc is already known, so we need two more pieces of information. AutoCAD will make an assumption (which you can reject) that the angle of the last line segment will be considered the direction (tangent) of the arc. If you accept this assumption, you should specify the endpoint. If not, use one of the following to specify the second piece of information:
  - The angle of the arc.
  - The center point of the arc.
  - Another direction of the arc.
  - The radius of the arc.
  - The second point, which can be any point on the parameter of the arc.
- Based on the information selected as the second point, AutoCAD will ask you to supply the third piece of information.

2.9 CONVERTING POLYLINES TO LINES AND ARCS AND VICE-VERSA

- This is a very essential technique that will allow you to convert any polylines to lines and arcs and convert lines and arcs to polylines.

2.9.1 Converting Polylines to Lines and Arcs

- The **Explode** command will explode a polyline into lines and arcs. To issue this command go to the **Home** tab, locate the **Modify** panel, then select the **Explode** button:

- AutoCAD will show the following prompt:

```
Select objects:
```

- Select the desired polylines and press [Enter] when done. The new shape will have lines and arcs.

2.9.2 Joining Lines and Arcs to Form a Polyline

- What we will discuss here is an option called **Join** within a command called **Edit Polyline**. To issue this command go to the **Home** tab, locate the **Modify** panel, then select the **Edit Polyline** button:

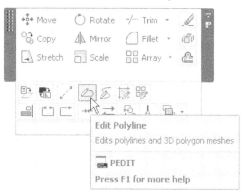

- You will see the following prompts:

```
Select polyline or [Multiple]:
Object selected is not a polyline
Do you want to turn it into one? <Y>
Enter an option [Close/Join/Width/Edit vertex/Fit/
Spline/Decurve/Ltype gen/Reverse/Undo]: J
```

- Start by selecting one of the lines or arcs you want to convert. AutoCAD will respond by telling you that the selected object is not a polyline! and giving you the option to convert this specific line or arc to a polyline. If you accept this, options will appear, and one of these options will be **Join**. Select the **Join** option, then select the rest of the lines and arcs. At the end, press [Enter] twice. The objects will be converted to a polyline.

DRAWING POLYLINES AND CONVERTING THEM

 Practice 2-5

1. Start AutoCAD 2012.
2. Open **Practice 2-5.dwg**.
3. Draw the following polyline using a start point of 18, 5 and width = 0.1.

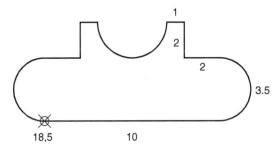

4. Then explode the polyline. As evidence, the width will disappear.
5. After exploding, the objects will be lines and arcs.
6. Save and close.

2.10 PRECISE DRAFTING USING OBJECT SNAP

- Object Snap, or OSNAP, is the most important accuracy tool in AutoCAD for 2D and 3D as well. It is a way to specify points on objects precisely using an AutoCAD database stored in the drawing file.

- ▪ Some of the Object Snaps include:
    - • **Endpoint**: To catch the Endpoint of an object (line, arc, or polyline).
    - • **Midpoint**: To catch the Midpoint of an object (line, arc, or polyline).
    - • **Intersection**: To catch the Intersection of two objects (any two objects).
    - • **Center**: To catch the Center of an object (arc, circle, or arcs in polylines).
    - • **Quadrat**: To catch the Quadrant of an object (arc, circle, or arcs in polylines).
    - • **Tangent**: To catch the Tangent of an object (arc, circle, or arcs in polylines).
    - • **Perpendicular**: To catch the Perpendicular point on an object (any object).
    - • **Nearest**: To catch a point on an object Nearest to your click point (any object).
- ▪ Here are some graphical presentations of each one of these OSNAPs:

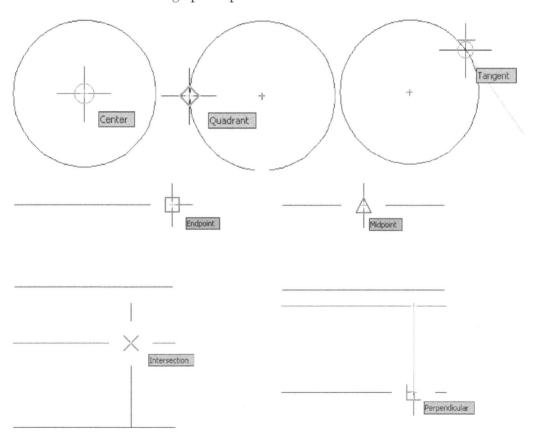

2.10.1 Activating Running OSNAPs

- To activate running OSNAPs in the drawing, click on the **Object Snap** button on the status bar:

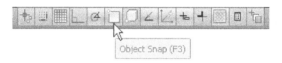

- When you switch this button on, you will use whatever OSNAPs were already on. In order to use your own, right-click the **Object Snap** button and the following menu will appear:

- You can switch on the desired Osnaps one-by-one.
- You can also select the **Settings** option and the following dialog box will appear:

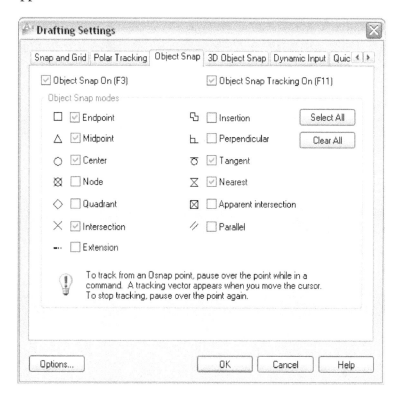

- There are two buttons at the right, **Select All** and **Clear All**. We recommend using Clear All and then selecting the desired OSNAPs. When done click **OK**.

2.10.2 OSNAP Override

- While the **Object Snap** button is on, some OSNAPs are working and others are not. Sometimes you may want to switch them all off and only use one; other times, you may want to switch everything back to normal. This is what we call Osnap Override.
- There are two ways to activate an override; they are:
  - Using the keyboard, type the first three letters of the desired OSNAP.
  - Using the keyboard, hold the [Shift] key and then right-click. You will see the following pop-up menu:

2.11 USING OBJECT SNAP TRACKING WITH OSNAP

- Sometimes OSNAP alone is not enough to specify desired points, especially if you need complex points. To solve this problem in the past we used to draw dummy objects to help us specify complex points. For example, when we wanted to specify the center of the circle at the center of a rectangle we used to a draw line from the midpoints of the two vertical lines, and the same for the horizontal lines. But since the introduction of Object Snap Tracking, or OTRACK, in AutoCAD 2000, drawing of dummy objects has no longer been needed. OTRACK depends on active OSNAP modes, which means if you want to use the midpoint with OTRACK, you have to switch the midpoint on first. To activate OTRACK go to the status bar and click the **Object Snap Tracking** button on:

- The procedure is very simple:
  - Using OSNAP go to the desired point and hover over it for a couple of seconds (don't click), then move to the right or left (also up and down depending on the next point), and you will see an infinite line extending in both directions (this line will be horizontal or vertical depending on your movement).
  - If you want to use a single point to specify your desired point, move in the desired direction, type in the desired distance, and press [Enter].
  - If you need two points, go the next point and hover for a couple of seconds, then move in the desired direction. Another infinite line will appear. Go to the intersection point of the two infinite lines and that will be your point.

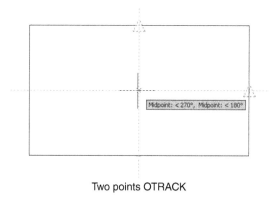

Two points OTRACK

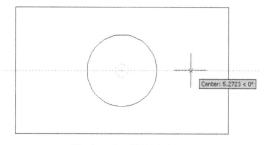

Single point OTRACK

- The Polar command is a major help here as well. Let's go back to the same dialog box:

- Under **Object Snap Tracking Settings**, there are two choices:
  - Track orthogonally only, the default option.
  - Track using all polar tracking settings.
- This means you can use the current polar angles (increment and additional angles) to specify points using OTRACK.
- To deactivate an OTRACK point, hover at the same point again for a couple of seconds and it will be deactivated.

DRAWING USING OSNAP AND OTRACK

 Practice 2-6

1. Start AutoCAD 2012.
2. Open **Practice 2-6.dwg**.
3. Using the proper OSNAP, create the four arcs as shown below:

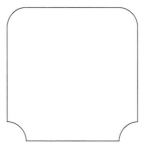

4. Then create the two circles as shown below (radius = 1).

5. Using OSNAP and OTRACK (using two points) draw the circle at the center of the shape (radius = 3.0).

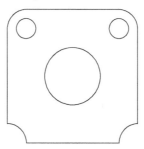

6. Using OSNAP and OTRACK (one point) draw the two circles at the right and left (distance center-to-center = 5.0 and radius = 0.5).

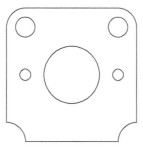

7. Change the increment angle in the Polar Tracking dialog box to 45. Make sure that **Track using all polar angle settings** is on and draw a circle (radius = 0.5), its center specified using OSNAP and OTRACK and Polar Tracking as shown below:

8. Use the same procedure to draw a circle at the top to end up with the final shape as shown here:

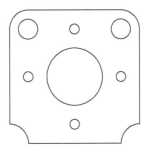

9. Save and close the file.

NOTES:

CHAPTER REVIEW

1. The Polyline command is different from the Line command because:
   a. It will produce lines and arcs
   b. All segments drawn using the same command are considered a single object
   c. You can specify a starting and ending width
   d. All of the above
2. Perpendicular, Tangent, and Endpoint are some of the _____ available in AutoCAD.
3. OTRACK can work by itself.
   a. True
   b. False
4. Which one of the following is not part of the eight pieces of information AutoCAD needs to draw an arc?
   a. Endpoint of the arc
   b. Midpoint of the arc
   c. Center point of the arc
   d. Angle
5. To convert lines and arcs from a polyline, use the _____ command.
6. If you want Grid to follow the Snap settings, set it up to be _____ in both X and Y.
7. Using the Polar Tracking dialog box, you can only track the orthogonal angles.
   a. True
   b. False
8. The _____ is always required to draw an arc.
9. To convert lines and arcs to polylines use the:
   a. Polyline Edit, Convert option
   b. Polyline Edit, Union option
   c. Polyline Edit, Join option
   d. None of the above

CHAPTER REVIEW ANSWERS

1. d
3. b
5. Explode
7. b
9. c

Chapter **3**

MODIFYING COMMANDS PART I

In This Chapter

◇ Different methods for selecting objects and selection cycling
◇ How to erase objects
◇ How to move and copy objects
◇ How to rotate and scale objects
◇ How to mirror, stretch, and break objects
◇ How to use grips for editing

3.1 HOW TO SELECT OBJECTS IN AUTOCAD

■ In order to use any of the modification commands we will discuss in this chapter you have to select the desired objects as the first step. Once you issue any of the modifying commands, the following prompt will appear:

```
Select objects:
```

■ The cursor will change to a pick box. At this prompt, you can work without typing anything on the keyboard, or you can type a few letters to activate a certain mode.
■ Without typing any letter, you can do the following things:
  • Select objects by clicking them one-by-one using the pick box.
  • Without selecting any object, click and move to the right to start **Window** mode, which will allow you to select all objects contained fully inside the window.
  • Without selecting any object, click and move to the left to start **Crossing** mode, which will allow you to select all objects contained fully inside the crossing, or touched (crossed) by it.

- See the following two examples:
- Window Example:

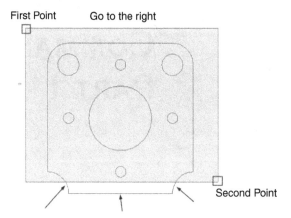

- All objects will be selected except the three objects with an arrow pointing to them. Why? Because either they are not fully contained inside the window or because they are not contained at all.
- Crossing Example:

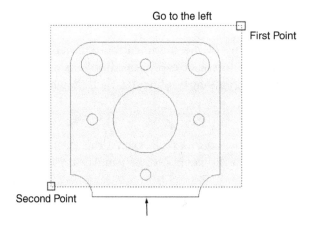

- All objects will be selected except the horizontal line with an arrow pointing to it. Why? Because it was neither contained nor crossed.
- Another way to use the Select objects prompt is to type a few letters to activate a certain mode. These letters are discussed as follows.

3.1.1 Window Mode (W)

- At the command prompt, type **W** and you will switch the selecting mode to **Window,** which will be available whether you go to the right or to the left.

3.1.2 Crossing Mode (C)

- At the command prompt, type **C** and you will switch the selecting mode to **Crossing,** which will be available whether you go to the right or to the left.

3.1.3 Window Polygon Mode (WP)

- **WP** mode will allow you to specify a non-rectangular window by specifying points in any fashion you like. By typing WP and pressing [Enter], the following prompt will appear:

```
First polygon point:
Specify endpoint of line or [Undo]:
Specify endpoint of line or [Undo]:
```

- Press [Enter] to end WP mode. WP is just like W; it needs to contain the object fully in order to select it.
- See the following example:

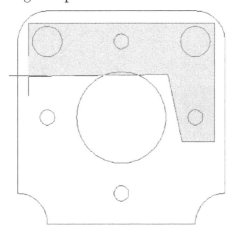

- Only four circles will be selected, and they are fully contained inside the WP.

3.1.4 Crossing Polygon Mode (CP)

- Since there is W and WP, it is obvious that there is also **C** and **CP**. Crossing Polygon is just like WP; it will allow you to specify a non-rectangular shape to contain and cross objects. See the following illustration:

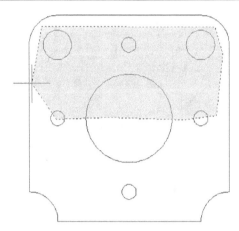

■ Three circles will be selected because they are fully contained inside the CP, and three more will be selected because they are crossed by the CP.

3.1.5 Fence Mode (F)

■ **Fence** mode will allow you to select multiple objects by crossing (touching) them. The lines of the fence can cross, contrary to WP and CP. See the following example:

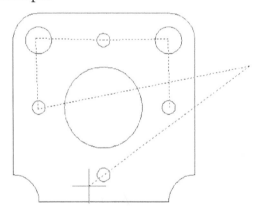

■ All the circles are selected because they are touched by the fence. The vertical line at the right will be selected as well.

3.1.6 Last (L), Previous (P), and All Modes

■ You can also select objects using the following modes:
   • **Last (L)**: To select the last object drawn.
   • **Previous (P)**: To select the last selected objects.
   • **All**: To select all objects in the drawing.

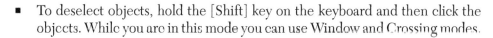 ▪ To deselect objects, hold the [Shift] key on the keyboard and then click the objects. While you are in this mode you can use Window and Crossing modes.

3.1.7 Other Methods to Select Objects

▪ There are two ways to use Modifying commands. You can issue the command, then select objects, or you can select objects, then issue the command. This technique is called the **Noun/Verb** technique. Without issuing any command, you can:
  • Select a single object by clicking it.
  • Click an empty space and go the right to get to **Window** mode.
  • Click an empty space and go the left to get to **Crossing** mode.
  • Click an empty space, then type **W** to get to **Window Polygon** mode.
  • Click an empty space, then type **C** to get to **Crossing Polygon** mode.
  • Click an empty space, then type **F** to get to **Fence** mode.
▪ When you select objects, you can go to the **Home** tab, locate the **Modify** panel, and issue the desired command, or you can right-click and you will get a shortcut menu that contains five modifying commands: Erase, Move, Copy Selection, Scale, and Rotate, as shown below:

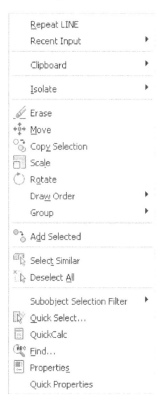

- By default, **Noun/Verb** is on, but to change it, take the following steps:
  - Go to the **Application Menu** and select the **Options** button.
  - Select the **Selection** tab.
  - Under **Selection modes**, make sure that **Noun/Verb** selection is on.

- Use this dialog box as well to do the following:
  - Turn on **Allow press and drag on object** to create a Window or Crossing even if the cursor is not over a clean spot for picking.
  - Under **Window selection method**, make sure that **Both Automatic detection** is selected so if you click then release the mouse, or you click and drag, both methods will be accepted.

3.2 SELECTION CYCLING

- While you are drafting using AutoCAD you may unintentionally draft objects over each other. Or, you may click on an object using a point that is shared by other objects. This issue used to be a problem in the past, but not anymore. The **Selection Cycling** feature in AutoCAD will notify you if your click touches more than one object and will give you the chance to choose the desired one. To activate **Selection Cycling** (by default it is active) go to the status bar and click the button as shown:

- After activating it and clicking on one object, you may see the following:

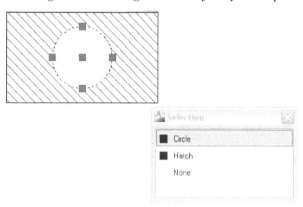

- The small window tells you that there are two possible objects, Circle and Hatch, and the Circle is selected. Using this window, if you move the mouse to Hatch, you will get the following:

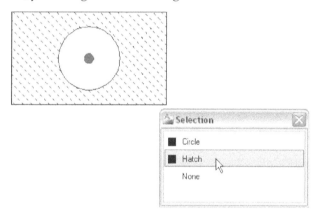

3.3 ERASE COMMAND

- The **Erase** command will delete any object you select. To issue the command go to the **Home** tab, locate the **Modify** panel, and then select the **Erase** button:

- You will see the following prompt in the Command window:

```
Select Objects:
```

- Select the desired objects. The Select Objects prompt is repetitive, so you always need to end it by pressing [Enter], or by right-clicking.
- You can also erase using other methods:
  - Click on the desired object(s) and then press the [Del] key on the keyboard.
  - Click on the desired object(s) and then right-click and a shortcut menu will appear; select **Erase**:

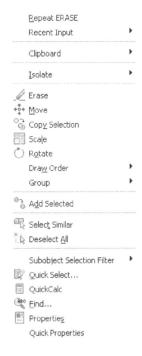

SELECTING OBJECTS AND THE ERASE COMMAND

 Practice 3-1

1. Start AutoCAD 2012.
2. Open **Practice 3-1.dwg**.

3. Start the Erase command (for all steps, after you finish selecting, press [Enter], then undo what you did).

4. Without typing any letter, find an empty space at the left, click (release your finger, don't hold the mouse button down) and go the right. Try to get the following result:

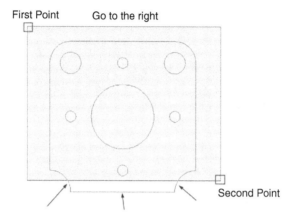

5. What was selected? Did you expect this result?
6. Do the same steps to simulate the following picture:

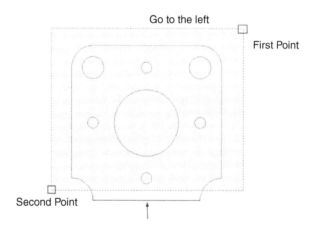

7. What was selected and why? _____

8. Try to simulate the following picture using WP:

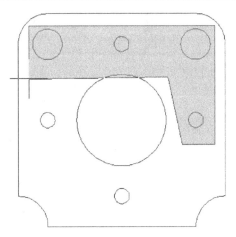

9. Try to simulate the following picture using CP:

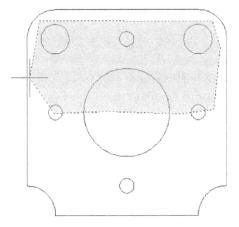

10. Try to simulate the following picture using Fence:

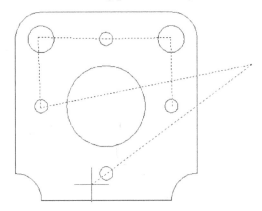

11. Without issuing the Erase command, select objects, then press [Del] on the keyboard to delete them. Undo what you did.
12. Without issuing the Erase command, select objects, then right-click and select the Erase command from the menu. Undo what you did.
13. Start the Erase command, select objects, then hold [Shift] to deselect some of the selected objects, and press [Enter]. Undo what you did.
14. Close the file without saving.

3.4 MOVE COMMAND

- This command will move objects from one place to another in the drawing. To issue this command, go to the **Home** tab, locate the **Modify** panel, then select the **Move** button:

- The following prompts will be shown:

```
Select objects:
Specify base point or [Displacement] <Displacement>:
Specify second point or <use first point as
displacement>:
```

- The first step is to select objects, and then you should select the base point. The base point is the point that will represent the objects; it will move a distance in an angle, and objects will follow. The main objective of the base point is accuracy. The last prompt will ask you to specify the second point, or the destination of your movement.
- See the following illustration:

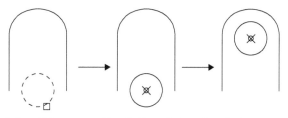

Select Objects   Specify Base Point   Specify Second Point

3.4.1 Nudge Functionality

- This function is very simple and will enable you to make an orthogonal move for selected objects. All you have to do is select objects, hold the [Ctrl] key on the keyboard, then use the four arrows on the keyboard, and you will see objects move in the desired direction.

MOVING OBJECTS

Practice 3-2

1. Start AutoCAD 2012.
2. Open **Practice 3-2.dwg**.
3. Using the Move command move the three circles and the rectangle to the correct places, so you get the following result (you will need OSNAP and OTRACK to move the rectangle accurately):

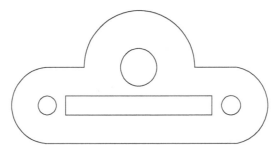

4. Save and close the file.

3.5 COPY COMMAND

- This command will copy objects. To issue this command go to the **Home** tab, locate the **Modify** panel, then select the **Copy** button:

- You will see the following prompts:

```
Select objects:
Current settings: Copy mode = Multiple
Specify base point or [Displacement/mode]
<Displacement>:
Specify second point or [Array] <use first point as
displacement>:
Specify second point or [Array/Exit/Undo] <Exit>:
Specify second point or [Array/Exit/Undo] <Exit>:
```

- After you select the desired objects, AutoCAD will report to you the current mode, in this case, Multiple. This mode allows you to create several copies in the same command. The other mode is Single copy. The first prompt will ask you to specify the base point. AutoCAD will then ask you to specify the second point to complete a single copy process, then repeats the prompt to create another one, and so on. There are three options you can use:
  - Undo, to undo the last copy.
  - Exit, to end the command.
  - Array, which will allow you to create an array of the same object using distance and angle. When you select the array option you will see the following prompts:

```
Enter number of items to array:
Specify second point or [Fit]:
```

- The first prompt is to input the number of items to array (including the original object); you will then specify the distance between the objects. You can use the **Fit** option to specify the total distance, and AutoCAD will equally divide the distance over the number of objects.
- See the following example of multiple copies:

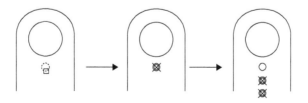

Select Objects   Specify Base Point Specify Second Point

COPYING OBJECTS

Practice 3-3

1. Start AutoCAD 2012.
2. Open **Practice 3-3.dwg**.
3. Copy the door using Multiple copying, and copy the toilet using the Array option (use Midpoint OSNAP for the toilet) to achieve the following:

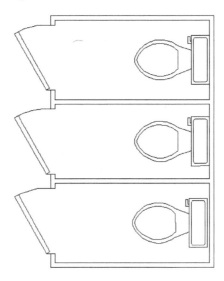

4. Save and close the file.

3.6 ROTATE COMMAND

- This command will rotate objects around the base point using a rotation angle or reference. To issue the command go to the **Home** tab, locate the **Modify** panel, and select the **Rotate** button:

- You will see the following prompts:

```
Current positive angle in UCS:
ANGDIR=counterclockwise ANGBASE=0
Select objects:
Select objects:
Specify base point:
Specify rotation angle or [Copy/Reference] <0>:
```

- The first message gives you the current angle direction and the angle base value. The base point here is the rotation point, which all the selected objects will rotate around.
- Use the **Copy** option to make a copy of the selected objects and then rotate them.
- See the following example:

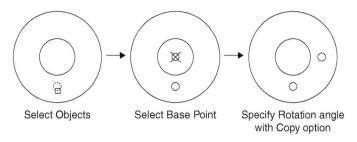

Select Objects   Select Base Point   Specify Rotation angle
                                                         with Copy option

3.6.1 Reference Option

- This option is very helpful if you don't know the rotation angle. You can specify two points to indicate the current angle and two points to input the new angle. You will see the following prompts:

```
Specify the reference angle <0>:
Specify second point:
Specify the new angle or [Points] <0>:
```

- You can input the angle by typing. If you want to input angles using two points, the first point you will pick will be for both angles; the current and the new. See the following illustration:

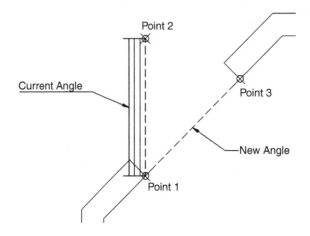

ROTATING OBJECTS

Practice 3-4

1. Start AutoCAD 2012.
2. Open **Practice 3-4.dwg**.
3. Rotate the lower window using Angle.
4. Rotate the upper window using Reference.
5. Rotate the chair with Copy mode to get the following result:

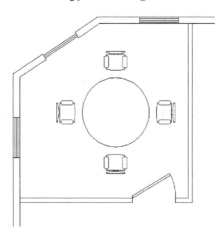

6. Save and close the file.

3.7 SCALE COMMAND

- This command will help you create larger or smaller objects using a scale factor or reference. To issue this command, go to the **Home** tab, locate the **Modify** panel, then select the **Scale** button:

- You will see the following prompts:

```
Select objects:
Specify base point:
Specify scale factor or [Copy/Reference] <1.0000>:
```

- The base point here is the scaling point; all the selected objects will be larger or smaller relative to it. Use the **Copy** option to make a copy of the selected objects and then scale them. See the following example:

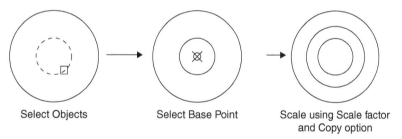

| Select Objects | Select Base Point | Scale using Scale factor and Copy option |

3.7.1 Reference Option

- This option is very handy if you don't know the scaling factor as a number; you can specify two points to indicate the current length and two points to input the new length. You will see the following prompts:

```
Specify reference length <0'-1">:
Specify second point:
Specify new length or [Points] <0'-1">:
```

- You can input length by typing. If you want to input lengths using two points, the first point you will pick will be for both lengths; the current and the new. See the following illustration:

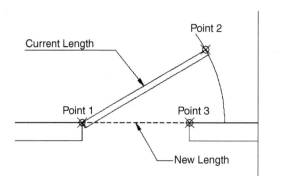

SCALING OBJECTS

Practice 3-5

1. Start AutoCAD 2012.
2. Open **Practice 3-5.dwg**.
3. Scale the toilet by scale factor = 0.9 using the midpoint of the wall.
4. Scale the sink by scale factor = 1.2 using the quadrant of the sink.
5. Scale the door using the Reference option to fit in the door opening.
6. You will get the following:

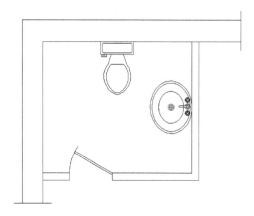

7. Save and close the file.

3.8 MIRROR COMMAND

- This command will help you create a mirror image of the selected objects using a mirror line. To issue this command, go to the **Home** tab, locate the **Modify** panel, then select the **Mirror** button:

- You will see the following prompts:

```
Select objects:
Specify first point of mirror line:
Specify second point of mirror line:
Erase source objects? [Yes/No] <N>:
```

- After selecting objects, specify the mirror line by specifying two points (you don't need to draw a line to be able to specify these two points). The last prompt will ask you to keep or delete the original objects.
- Text can be part of the selection set to be mirrored. You can tell AutoCAD what to do with it (copy or mirror) by using system variable MIRRTEXT. To issue this command, type it in the Command window, and you will see the following:

```
Enter new value for MIRRTEXT <0>:
```

- You can input either 1 (which means mirror the text just like the other objects) or 0 (which means copy the text).
- See the following example:

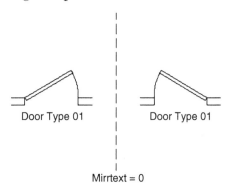

Door Type 01   Door Type 01

Mirrtext = 0

MIRRORING OBJECTS

Practice 3-6

1. Start AutoCAD 2012.
2. Open **Practice 3-6.dwg**.
3. Mirror the entrance door to open inside rather than outside, keeping the text as is.
4. Mirror all windows at the right to the left.
5. Mirror the furniture of the room at the right to the room at the left.
6. Mirror the two windows at the top (to the one at the right and the one in the middle) to the lower wall.
7. You should get the following result:

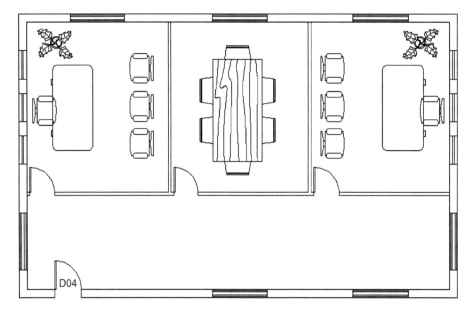

8. Save and close.

3.9 STRETCH COMMAND

- This command will allow you to change the length of selected objects by stretching using distance and angle. To issue this command, go to the **Home** tab, locate the **Modify** panel, and select the **Stretch** button:

- You will see the following prompts:

```
Select objects to stretch by crossing-window or
crossing-polygon...
Select objects:
Specify base point or [Displacement] <Displacement>:
Specify second point or <use first point as
displacement>:
```

- The Stretch command is different than other modifying commands we have learned about because it asks you to select the objects desired using C or CP modes. Why? Because Stretch will utilize both features of C and CP, containing and crossing. All objects contained fully inside the C or CP will be moving, while objects crossed will be stretched either by increasing or decreasing the length. You should then specify the base point (the same principle of the Move and Copy commands), and finally, specify the second point. See the following example:

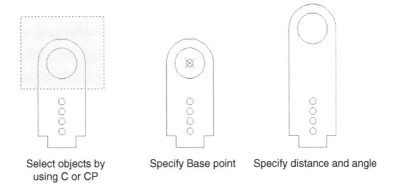

Select objects by   Specify Base point   Specify distance and angle
using C or CP

STRETCHING OBJECTS

Practice 3-7

1. Start AutoCAD 2012.
2. Open **Practice 3-7.dwg**.
3. The vertical distance of the three rooms is not correct; it should be 1'-0" more. Use the Stretch command to do that.
4. The door of the room at the right is positioned wrong; stretch it to the right, for 2'-5".
5. The door of the room at the left is also positioned wrong; stretch it to the right to the small line indicating the new position.
6. You should have the following:

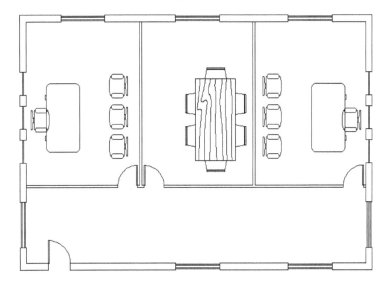

8. Save and close the file.

STRETCHING OBJECTS

Practice 3-8

1. Start AutoCAD 2012.
2. Open **Practice 3-8.dwg**.

3. Stretch the upper part to look like the lower part by using a distance = 1.00 and CP.

4. You should have the following:

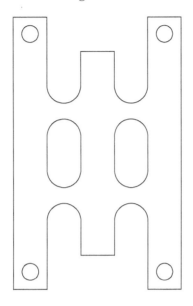

5. Save and close the file.

3.10 BREAK COMMAND

- This command will help you break any object into two objects by removing the portion between two specified points. To issue this command go to the **Home** tab, locate the **Modify** panel, and select the **Break** button:

- You will see the following prompts:

```
Select object:
Specify second break point or [First point]:
```

- If selecting the object also specifies the first point, then select the second point. This will end the command. But if you consider selecting is purely selecting, and you didn't specify the first point, then type **F**, and you will see the following two prompts:

```
Specify first break point:
Specify second break point:
```

- Using these two prompts, specify two points on the object and the Break command will end.
- AutoCAD offers the same command using a different technique, breaking on the same point. To issue this command, go to the **Home** tab, locate the **Modify** panel, then select the **Break at Point** button:

- You will see the following prompts:

```
Select object:
Specify second break point or [First point]: _f
Specify first break point:
Specify second break point: @
```

- After you select the desired object, it will flip automatically to allow you to specify the first point. At the second point prompt, AutoCAD responds with @, which means "use the same first point."
- If the object you are breaking is a circle, make sure to specify the two points CCW. See this example:

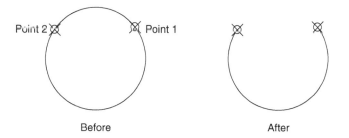

Before   After

BREAKING OBJECTS

Practice 3-9

1. Start AutoCAD 2012.
2. Open **Practice 3-9.dwg**.
3. In order to locate the meeting table exactly at the center of the room, horizontally and vertically, we need to break the upper horizontal line and the left vertical line.
4. Using the Break command, break the upper horizontal line from the points shown below:

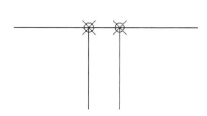

5. Using the Break command break the left vertical line from the points shown below:

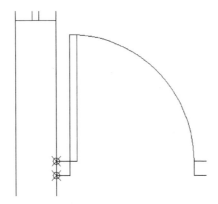

6. Using the Move command, move the meeting table to the center of the room using OSNAP and OTRACK.
7. You will get the following:

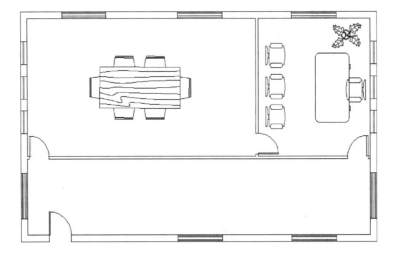

8. Save and close the file.

3.11 USING GRIPS TO EDIT OBJECTS

- **Grips** are blue squares and rectangles that appear on objects and enable you to modify them using five modifying commands using the grips as the base point. It is a clever tool for performing modifying tasks faster without compromising the accuracy of the conventional modifying commands discussed earlier.

■ Depending on the type of object, grips will appear in different places. See the following illustration:

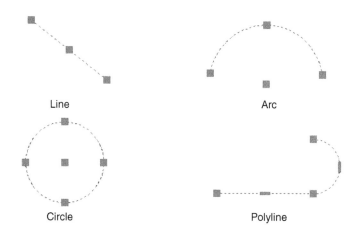

■ When you click an object, it will be selected, and you can issue any of the modifying commands we have discussed. But if you click any of the blue squares (grips) the object will turn red, which means it will become the active base point (hot) for the following five modifying commands:
   • Stretch
   • Move
   • Rotate
   • Scale
   • Mirror

■ Sometimes you can use the Lengthen command (discussed in the next chapter) depending on which grip was clicked and which object was selected. You can see these five commands by right-clicking:

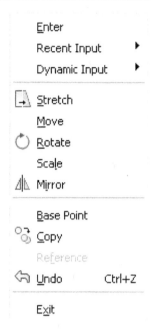

- These commands have one thing in common, the base point, except for Mirror. So why is Mirror in the list? And why isn't Copy among the commands in the list? To answer the first question, AutoCAD considers the first point of the mirror line as the base point, so it is included in the list. To answer the second question, Copy here is a mode and not a command, which means it will work with all five commands.
- Holding [Shift] while selecting the grip will enable you to select more than one base point.
- Holding [Ctrl] while specifying any of the following, the second point (Move and Stretch), rotation angle (Rotate), scale factor (Scale), second point of the mirror line (Mirror), will allow AutoCAD to remember the last input and repeat it graphically.
- The Base Point option will enable you to select another base point other than the grip selected.
- Let's look at an example. Select the following shapes (two polylines):

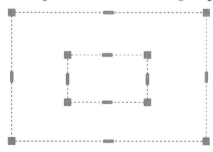

- Click the upper-right corner to make it hot, then right-click to access the menu, and select the Rotate command:

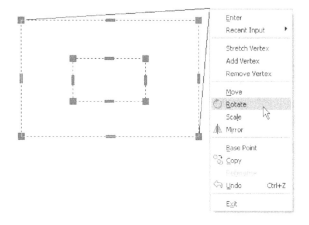

- Now you can rotate the two shapes around the grip (considered your base point); right-click again and select the Base Point option to select another base point, which will be the center of the two rectangles (using OSNAP and OTRACK):

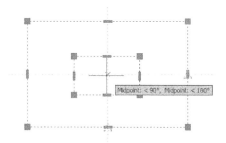

- Right-click again and select the Copy option to copy while rotating. Using Polar Tracking specify an angle of 90, then press [Esc]. You will get the following result:

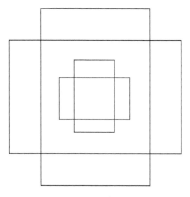

- Specific objects such as polylines will show more than the blue squares; they will show rectangles at the midpoint of each line and arc segment. The rectangles at the midpoint of the polyline have several functions. Depending on whether it is a line segment or arc segment, AutoCAD will show a different menu. Go to the grip and hover over it for a second (don't click) and you will see something like the following:

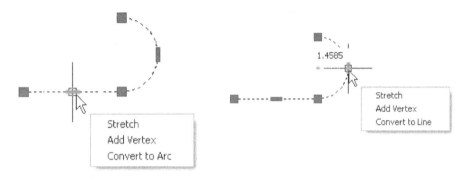

- As you can see, you can stretch the selected segment, add a vertex, convert a line to an arc, or convert an arc to line.
- When you are done with grips click [Esc] once or twice, depending on the situation you are in, and this will end the Grips mode.

3.12 GRIPS AND DYNAMIC INPUT

- If you stay on one of the grips (without clicking) Dynamic Input along with grips will help you get information about the selected objects based on their type. See the following examples:
- Using Line and one of the endpoints, you will see the length and angle with East, along with two commands, Stretch and Lengthen:

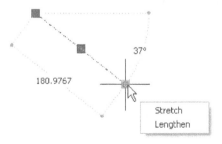

- The two connected lines will show two lengths and two angles:

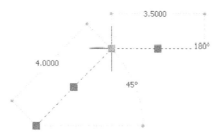

- The midpoint of an arc will show the radius and included angle, along with two commands, Stretch and Radius:

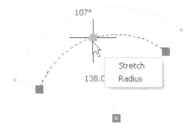

- However, the endpoints of an arc will show the radius and angle with the East, along with the Stretch and Lengthen commands:

- As for a circle, you will only see the radius:

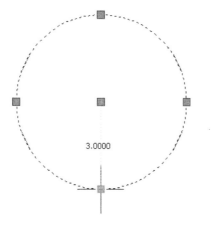

- The endpoint of a segment of a polyline will show the length of the shared lines, along with a shortcut menu that includes options such as Stretch Vertex, Add Vertex, and Remove Vertex:

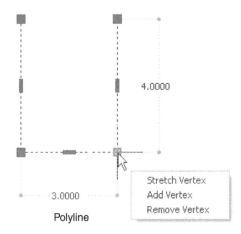

3.13 GRIPS AND PERPENDICULAR AND TANGENT OSNAPS

- Using grips you can specify perpendicular and tangent OSNAPs, keeping in mind that those two settings are turned on when running OSNAP. See the following two examples:
- Tangent Example:

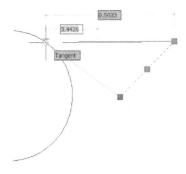

Line tangent to an arc

- Perpendicular Example:

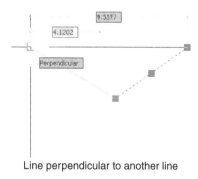

Line perpendicular to another line

USING GRIPS TO EDIT OBJECTS

Practice 3-10

1. Start AutoCAD 2012.
2. Open **Practice 3-10.dwg** (you should complete this practice using grips techniques only).
3. Select the large circle, make the center hot, and scale it with copying, with 1.2 as the scale factor.
4. Mirror the three lines at the right to the other side (you should use another base point, along with Copy mode).
5. You should have the following shape:

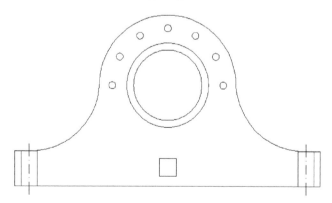

6. Select the polyline below the two big circles. Convert the top horizontal line to an arc by moving up a distance = 0.2 (this is not the new radius).

7. Using the rectangle grip at the middle of the lower horizontal line of the polyline stretch the line downward by distance = 0.1.

8. Using grips and Dynamic Input what is the horizontal distance and the vertical distance of the polyline? _____, _____

9. Press [Esc] to clear grips, then re-select the polyline again, select one of the grips (any one), and then right-click and select the Move command, then right-click again, and select Copy mode (make sure that Polar Tracking is on to help you get exact angles). Now move to the right, type 1 as a distance, but before you press [Enter] hold the [Ctrl] key so AutoCAD will remember this distance. Now while you are still holding the [Ctrl] key, make three copies to the right and three copies to the left.

10. The shape should look like the following:

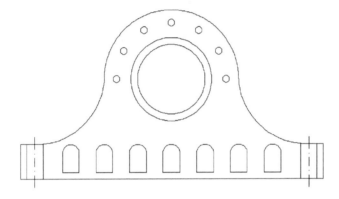

11. Save and close the file.

NOTES:

CHAPTER REVIEW

1. Which one of the following is not true about the Reference option?
   a. You can use it with Rotate and Scale
   b. In Rotate you have to specify the current and the new angle by specifying four points
   c. In Scale you have to specify two distances, the current and the new
   d. You can't use it with the Stretch command
2. The Stretch command will ask for something that other modifying commands will not.
   a. True
   b. False
3. If you want to select the last selected set, type _____ at the Command window.
4. With the Mirror command you should draw a line to act as a mirror line.
   a. True
   b. False
5. While using grips, holding _____ while selecting the grip will enable you to select more than one base point.
6. _____ will help you control the outcome of text in the Mirror command.
7. In grips there are _____ modifying commands in the right-click menu.
   a. Seven
   b. Six
   c. Five
   d. Four
8. Using grips, stretching the arrow at the midpoint of an arc will increase/decrease the _____ of an arc.
9. While using the Break command you can break at the same point using which character?
   a. $
   b. #
   c. @
   d. &

CHAPTER REVIEW ANSWERS

1. b
3. P
5. [Shift]
7. Five
9. c

MODIFYING COMMANDS PART II

Chapter **4**

In This Chapter

◇ How to offset objects
◇ How to fillet and chamfer objects
◇ How to trim and extend objects
◇ How to array objects
◇ How to lengthen and join objects

4.1 INTRODUCTION

■ The commands discussed in this chapter are modifying commands with special abilities. They can build over the shapes you draw. Each one of them has a unique function; two of them can create objects (Offset and Array), and the others can change an existing object's shape (the Fillet, Chamfer, Trim, Extend, Lengthen, and Join commands).

■ Here is a brief description of each command:
   • **Offset** command: to create copies of an object parallel to the original.
   • **Fillet** command: to create a neat intersection between two objects either by extending/trimming lines or using arcs.
   • **Chamfer** command: to create a neat intersection but only for lines. The neat intersection is created by extending/trimming the two lines or by creating a new line showing the chamfered edge.
   • **Trim** command: to trim objects using other objects as cutting edges.
   • **Extend** command: to extend objects using other objects as boundary edges.
   • **Array** command: to create objects using three different methods: rectangular, circular, and using a path.

- **Lengthen** command: to increase/decrease the length of a line/arc/polyline.
- **Join** command: to join lines to become a single line and join arcs to become a single arc, or to form a circle out of an arc.

4.2 OFFSETTING OBJECTS

- The Offset command will create copies of an object parallel to the original. The new object will possess the same properties of the original object. You can offset using offset distance, or using a point the new object will pass through. You can start this command by going to the **Home** tab, locating the **Modify** panel, then selecting the **Offset** button:

- You will see the following AutoCAD prompts:

```
Current settings: Erase source=No Layer=Source OFFSETGAP-
TYPE=0
Specify offset distance or [Through/Erase/Layer] <Through>:
```

4.2.1 Offsetting Using the Offset Distance Option

- If you know the distance between the object and the new parallel copy, then input this value, select the original object, and then click on the side you want the new object to go on. You will see the following prompt:

```
Specify offset distance or [Through/Erase/Layer] <Through>:
Select object to offset or [Exit/Undo] <Exit>:
Specify point on side to offset or [Exit/Multiple/Undo]?
<Exit>:
```

- This will enable you to create a single offset. To create more offsets using the same offset command, select another object, and follow the same steps again. Pressing [Enter] or right-clicking will end the command.

4.2.2 Offsetting Using the Through Option

■ If you don't know the offset distance, but you know a point in the drawing the new parallel object will pass through, this option will help you accomplish your mission. You will see the following AutoCAD prompts:

```
Specify offset distance or [Through/Erase/Layer] <Through>:
Select object to offset or [Exit/Undo] <Exit>:
Specify through point or [Exit/Multiple/Undo] <Exit>:
```

■ This will enable you to create a single offset. To create more offsets using the same offset command, select another object and follow the same steps again. Pressing [Enter] or right-clicking will end the command.
■ Here is an example:

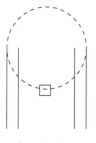 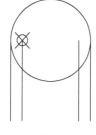 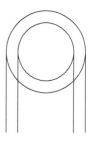

Select object   Specify Through point   The result

4.2.3 Using the Multiple Option

■ You can use the Multiple option to repeat the offset distance or Through option in the same command repeatedly by clicking on the side of the offset, or by specifying a new through point. The prompts for the Multiple option are:

```
Specify through point or [Exit/Multiple/Undo] <Exit>:
Specify point on side to offset or [Exit/Undo] <next
object>:
```

■ While offsetting keep the following in mind:
  • If you make a mistake use the Undo option.
  • AutoCAD remembers the last offset distance used and will save it in the file.
  • When offsetting an arc or circle, the new arc and circle will share the same center point.
  • If you offset a closed polyline, the output will be smaller or larger.

OFFSETTING OBJECTS

 Practice 4-1

1. Start AutoCAD 2012.
2. Open **Practice 4-1.dwg**.
3. Offset the circle to the inside by distance = 0.5.
4. Offset the rightmost vertical line to the inside using the Through option and the midpoint of the small horizontal line. Do the same for the left side. See the following illustration:

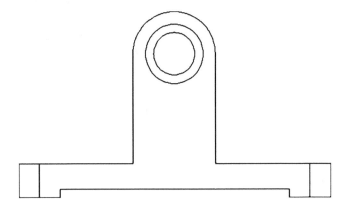

5. Using the two newly created lines offset each line to its right and its left by distance = 0.5.
6. You should have the following:

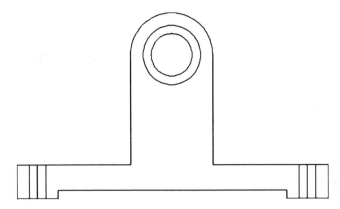

7. Save and close the file.

OFFSETTING OBJECTS

 Practice 4-2

1. Start AutoCAD 2012.
2. Open **Practice 4-2.dwg**.
3. Offset the polyline that represents the outside edge of the outer wall by distance = 1'-0".
4. Explode the inner polyline.
5. Offset the yellow line representing one of the stair steps 10 times using distance = 1'-6" and the Multiple option.
6. Offset the vertical line at the right to the left using the Through option to pass through the inner right end point of the arc.
7. Offset the newly created line to the right by distance = 6".
8. You will get the following:

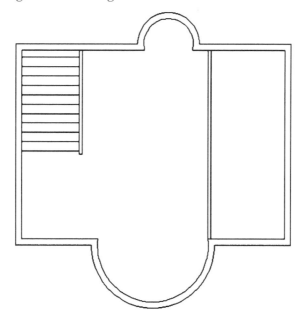

9. Save and close the file.

4.3 FILLETING OBJECTS

- The mission of the Fillet command is to create neat intersections. You should set the value of the radius as the first step. If it is 0 (zero) then you can only use Fillet between two lines and the command will extend/trim the lines to the proposed intersection point. But if the value of the radius is greater than 0 (zero) then Fillet can use lines and circles to fillet these objects with an arc. See the following:

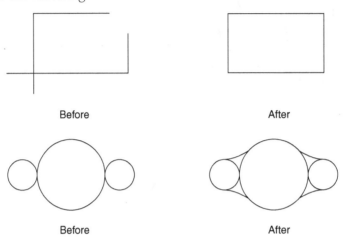

Before   After

Before   After

- While working with a radius greater than zero, you can select between **Trim**, which will allow you to trim the original objects or **No trim**, which will allow the original objects to stay as is. In both cases, you will be able to see the arc when hovering over the second object, and you can make sure that it is the right value!
- Here is an example:

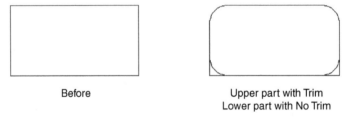

Before   Upper part with Trim
                                            Lower part with No Trim

- To start this command go to the **Home** tab, locate the **Modify** panel, then select the **Fillet** button:

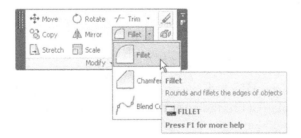

- You will see the following AutoCAD prompts:

```
Current settings: Mode = TRIM, Radius = 0.0000
Select first object or [Undo/Polyline/Radius/Trim/
Multiple]:
```

- You should always check the first line of this command because it will report the current value of Radius, and you can decide whether to keep it or change it. Type **r** or right-click and select the Radius option to set the new value of the radius. You will see the following prompt:

```
Specify fillet radius <0.0000>:
```

- Type **t** or right-click and select **Trim** to change the mode to **Trim** or **No trim**. You will see the following prompt:

```
Enter Trim mode option [Trim/No trim] <Trim>:
```

- The Fillet command will allow only a single fillet per command. To make multiple fillets in the same command, simply change the value to **Multiple**. If you make a mistake use the **Undo** option to undo the last action. To end the command press [Enter].
- You can do two important things while using the Fillet command:
  - You can fillet two parallel lines regardless of the current radius value.
  - You can fillet any two lines with radius = 0 regardless of the current value of radius by holding the [Shift] key.
- If you use the Multiple option you can use different radius values in the same command.

FILLETING OBJECTS

Practice 4-3

1. Start AutoCAD 2012.
2. Open **Practice 4-3.dwg**.
3. Using the Fillet command try to get the following final result:

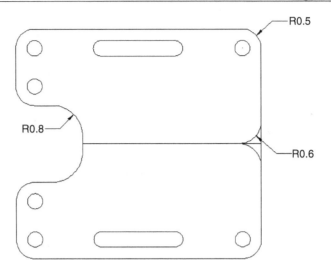

4. Save and close the file.

4.4 CHAMFERING OBJECTS

- The Chamfer command will create neat intersections as well, but with this command you should set the value of Distance (or Distance and Angle) as the first step. If it is 0 (zero) then you can use Chamfer to extend/trim the lines to the proposed intersection point. But if the value of Distance is greater than 0 (zero) then Chamfer will create a sloped edge between the two lines. While working with a distance greater than zero, you can select between **Trim**, which will allow you to trim the original objects or **No trim**, which will allow the original objects to stay as is. In both cases, you will be able to see the chamfer line when hovering over the second object, and you can make sure that it's the right value!
- To create the sloped edge, use one of the following two methods:
  - Distance (two distances)
  - Distance and Angle

4.4.1 Chamfering Using the Distance Option

- You will see the following two prompts when you chamfer using the Distance option:

```
Specify first chamfer distance <0.0000>:
Specify second chamfer distance <0.0000>:
```

- There will be two different cases; they are:

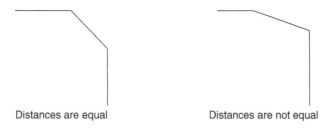

Distances are equal   Distances are not equal

4.4.2 Chamfering Using the Distance and Angle Options

- You will see the following two prompts when you chamfer using Distance and Angle:

```
Specify chamfer length on the first line <0.0000>:
Specify chamfer angle from the first line <0>:
```

- Set the length (which will be cut from the first object selected) and an angle.
- See the following illustration:

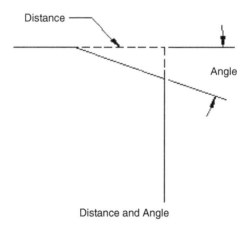

Distance and Angle

- To start this command go to the **Home** tab, locate the **Modify** panel, then select the **Chamfer** button:

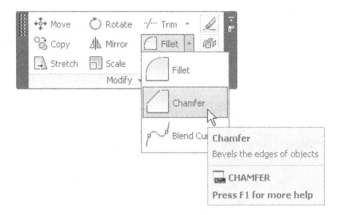

- You will see the following prompts:

```
(TRIM mode) Current chamfer Dist1 = 0.0000, Dist2 =
0.0000
Select first line or [Undo/Polyline/Distance/Angle
/Trim/method/Multiple]:
```

- You should always check the first line of this command because it will report the current value of the method used (whether Distance or Distance and Angle) and the current values. Accordingly, you can decide to keep it or change it.
- The other options, Multiple, Trim, and Undo, are identical to what we learned for the Fillet command. The Method option is used to select the default method to be used in the chamfering process.
- Another similarity to the Fillet command is you have to hold the [Shift] key, which will allow you to chamfer by extending/trimming the two lines, regardless of the current distance values.
- Trim and Untrim in Chamfer will affect the Fillet command and vice-versa.

CHAMFERING OBJECTS

 Practice 4-4

1. Start AutoCAD 2012.
2. Open **Practice 4-4.dwg**.
3. Use the Chamfer command to make the shape look like the following:

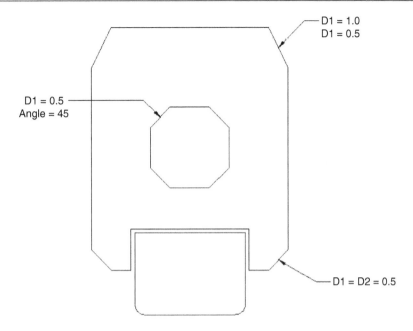

D1 = 1.0
D1 = 0.5

D1 = 0.5
Angle = 45

D1 = D2 = 0.5

4. Save and close the file.

4.5 TRIMMING OBJECTS

- This command will allow you to remove part of an object based on cutting edge(s). In order to successfully complete this command, you have to select the cutting edges first, press [Enter], and then select the part of the objects you want to remove.
- The following example illustrates the process of trimming:

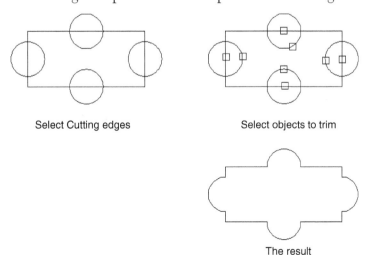

Select Cutting edges   Select objects to trim

The result

- To issue this command go to the **Home** tab, locate the **Modify** panel, then select the **Trim** button:

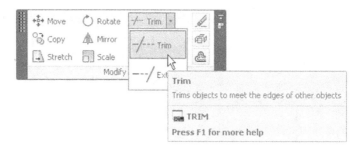

- You will see the following prompt:

```
Current settings: Projection=UCS, Edge=Extend
Select cutting edges ...
Select objects or <select all>:
```

- The first line is a message from AutoCAD telling you the current settings. The second line asks you to select the cutting edges, which is the first step of the Trim command. When you're done press [Enter] or right-click. You can also use the Select All option, which will select all the objects as the cutting edges. This can be done by pressing [Enter].
- You will see the following prompt either way:

```
Select object to trim or shift-select to extend or
[Fence/Crossing/Project/Edge/eRase/Undo]:
```

- You can start by clicking on the parts of the objects you want to remove. You have three ways to do that:
  - Select the objects one by one.
  - Click on an empty space and go to the left to start Crossing mode to select the objects to be removed collectively.
  - Type F to start the Fence option (discussed in Chapter 3).
- If you make a mistake simply type u to undo the last trim.
- While you are trimming you may end up with some orphan objects, but you can get rid of them with the eRase option (type r).

TRIMMING OBJECTS

 Practice 4-5

1. Start AutoCAD 2012.
2. Open **Practice 4-5.dwg**.
3. Using the Trim command try to make the shapes look like the following:

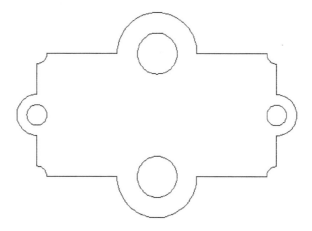

4. Save and close the file.

4.6 EXTENDING OBJECTS

- This command will allow you to extend the end of an object to the boundary edge(s). In order to successfully complete this command, you have to select the boundary edges first, press [Enter], and then select the ends of the objects you want to extend.
- See the following illustration:

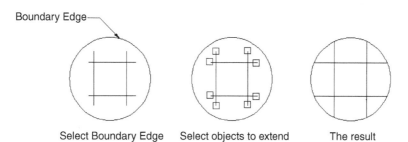

Boundary Edge

Select Boundary Edge   Select objects to extend   The result

■ To issue this command go to the **Home** tab, locate the **Modify** panel, then select the **Extend** button:

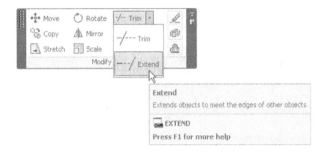

■ You will see the following prompts:

```
Current settings: Projection=UCS, Edge=Extend
Select boundary edges ...
Select objects or <select all>:
```

■ The first line is a message from AutoCAD telling you the current settings. The second line asks you to select the boundary edges, which is the first step of the Extend command. When you're done press [Enter] or right-click. You can also use the Select All option, which will select all the objects as the boundary edges. This can be done by pressing [Enter]. Either way you will see the following prompt:

```
Select object to extend or shift-select to trim or
[Fence/Crossing/Project/Edge/Undo]:
```

■ You can start clicking on the parts of the objects you want to extend. You can do this one of three ways:
  • Select the object ends one by one.
  • Click an empty space and go to the left to start Crossing mode to select the objects to be removed collectively.
  • Type F to start the Fence option (discussed in Chapter 3).
■ If you make a mistake simply type u to undo the last extend.
■ The last feature in both the Trim and Extend commands is the ability to use each command while you are using the other. See the following example:

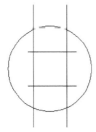 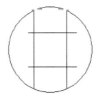 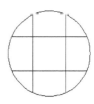

Select the circle as Cutting Edges   Trim lines as shown   Hold [Shift] to convert Trim
command to Extend command
and select other lines

EXTENDING OBJECTS

 Practice 4-6

1. Start AutoCAD 2012.
2. Open **Practice 4-6.dwg**.
3. Using the Extend command (and Trim if needed) correct the architectural plan so it looks like the following:

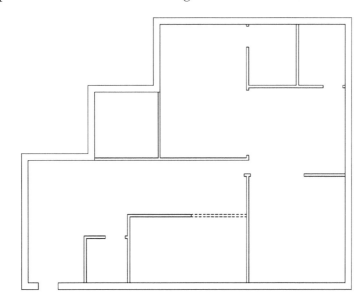

4. Save and close.

4.7 ARRAYING OBJECTS

- Array commands help you create several duplicates of objects using three methods:
  - Rectangular array
  - Path array
  - Polar array

4.7.1 Rectangular Array

- This command will allow you to create replicates in a matrix fashion using rows and columns. To issue this command go to the **Home** tab, select the **Modify** panel, and select the **Rectangular Array** button:

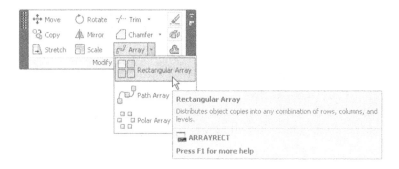

- The following prompts will be shown:

```
Select objects: 1 found
Select objects:
Type = Rectangular Associative = Yes
Specify opposite corner for number of items or [Base
point/Angle/Count] <Count>:
```

- The first prompt will ask you to select the desired objects. When you're done press [Enter]. The second prompt will show that the type of array is rectangular and that associativity is on. Assuming the location of the selected object, AutoCAD will ask you to specify a window and provide two details: the number of rows and the number of columns. If you don't want to specify a window, then select the Count option by pressing [Enter], and you will see the following two prompts:

```
Enter number of rows or [Expression] <4>:
Enter number of columns or [Expression] <4>:
```

- Type in the number of rows and columns. When you press [Enter] you will see the following prompt:

```
Specify opposite corner to space items or [Spacing]
<Spacing>:
```

- You are asked to specify the spacing between the rows and columns by specifying a window. The movement of the cursor will specify whether the distance will be positive or negative in both directions. If you don't want to do that, then select the option Spacing, and you will see the following two prompts:

```
Specify the distance between rows or [Expression]:
Specify the distance between columns or [Expression]:
```

- Type in the distances between the rows and columns, keeping in mind that if you want to create duplicates down and left you should use the minus sign. Press [Enter] to end the command.
- Once the Array command is finished, because it's associative, when you select any object of the array you will see something like the following:

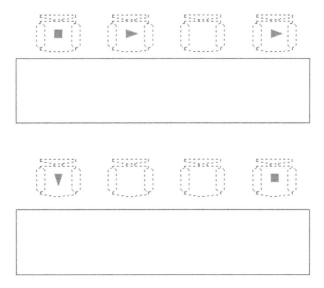

- Each one of these has a function.

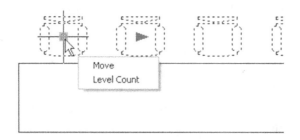

- This grip is used for moving all the array objects or to control Level Count (3D only). Use [Ctrl] to browse between these options.

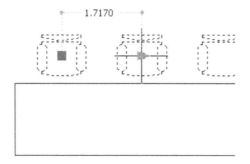

- This grip is for changing the column spacing.

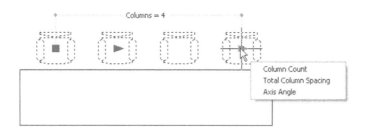

- This grip is for changing Column Count, Total Column Spacing, or Axis Angle (to specify another angle other than horizontal and vertical). Use [Ctrl] to browse between these options.

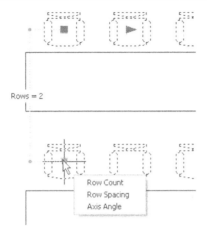

- This grip is for changing Row Count, Row Spacing, or Axis Angle. Use [Ctrl] to browse between these options.

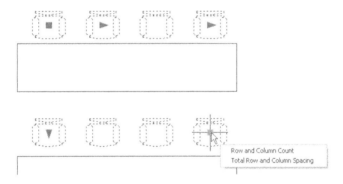

- This grip is for changing Row and Column Count or Total Row and Column Spacing. Use [Ctrl] to browse between these options.
- Meanwhile, a context tab named Array will appear; it looks something like the following:

- This tab will help you modify any of the values by typing them in.
- Levels is for Array in 3D.

ARRAYING OBJECTS USING RECTANGULAR ARRAY

 Practice 4-7

1. Start AutoCAD 2012.
2. Open **Practice 4-7.dwg**.
3. Using the chair try to create a rectangular array like the following:

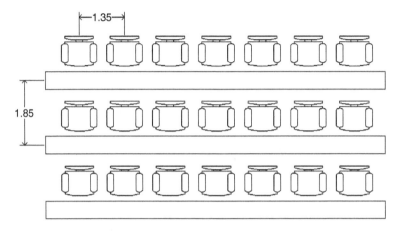

4. Save and close the file.

4.7.2 Path Array

- This command will allow you to create an array using an object such as a polyline, spline, arc, etc. To issue this command go to the **Home** tab, locate the **Modify** panel, and select the **Path Array** button:

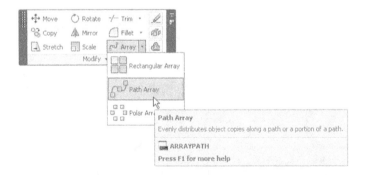

- The following prompts will be shown:

```
Select objects:
Type = Path  Associative = Yes
Select path curve:
Enter number of items along path or
[Orientation/Expression] <Orientation>:
```

- The first prompt will ask you to select the desired objects; press [Enter] when you're done. The second prompt gives the type of array and path and shows that associativity is on. AutoCAD then asks you to specify the number of items along the path; you can type the desired number or move the mouse along the path to indicate the number. Using this method, the objects to be arrayed will not be aligned with the path automatically, but you can align them manually. You can skip specifying the number of items by selecting the Orientation option. The following prompts will be shown:

```
Specify base point or [Key point] <end of path
curve>:
Specify direction to align with path or
[2Points/NORmal] <current>:
```

- AutoCAD asks you to specify a base point for the object. You can either specify the direction of the array or use the 2-Points option or Normal option (this option is for 3D). See the following example.
- Assume we have an arc with a chair like the following:

- Note the current location of the chair relating to the arc. Using the Orientation option, specify a base point and two points like the following:

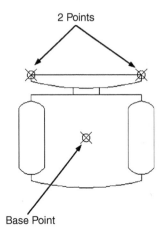

- This will be the result:

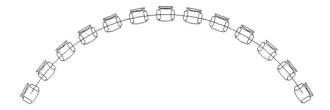

- The following prompts will be shown:

```
Enter number of items along path or [Expression] <4>:
Specify the distance between items along path or
[Divide/Total/Expression] <Divide evenly along path>:
```

- You are asked to specify the number of items along the path. You can do this by typing the desired number or moving the mouse along the drawn path. The final step is to specify the distance between items along the path. The spacing along with the number of items should be equal or less than the total length of the path. You can also:
  - Divide, i.e., divide evenly along the path (this is the default option).
  - Total, and you will see the following prompt:

```
Enter total distance between start and end items:
```

- Press [Enter] to end the command. Once the array command is finished, because it's associative, when you select any object of the array you will see something like the following (these grips will appear only if you selected Total as the distance method):

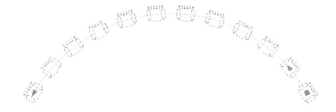

- The three grips are for:

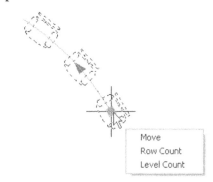

- The grip at the right is for moving the whole array or changing the Row Count or Level Count (3D only). For Row Count, see the following illustration, which includes three rows:

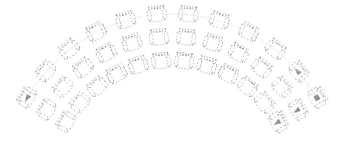

- Two additional grips are used to control the distance between rows and the total number of rows.
- The second grip is for item spacing:

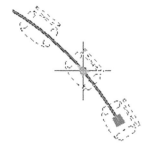

- The third grip is:

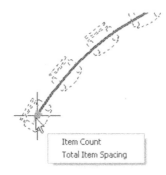

- This grip is for changing the Item Count and Total Item Spacing.
- Meanwhile, a new context tab named Array will appear; it looks like the following:

- This tab will help you modify any of the values by typing them in.
- Levels is for Array in 3D.

ARRAYING OBJECTS USING THE PATH ARRAY

 Practice 4-8

1. Start AutoCAD 2012.
2. Open **Practice 4-8.dwg**.
3. Using the Path Array try to create an array, keeping the following in mind:
   a. Number of items = 22
   b. Equally spaced
   c. Number of rows = 2
   d. Distance between rows = 1.25
4. You should use the Orientation option, specifying:
   a. Base point = center of chair (using OSNAP and OTRACK)
   b. 2 Points = left point of the top part, then right point of the top part
5. Erase the outer polyline.
6. You should get the following result:

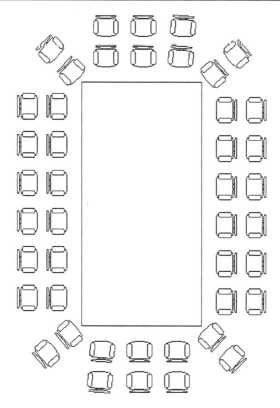

7. Save and close the file.

4.7.3 Polar Array

- This command will allow you to duplicate objects in a circular fashion. To issue this command go to the **Home** tab, locate the **Modify** panel, then select the **Polar Array** button:

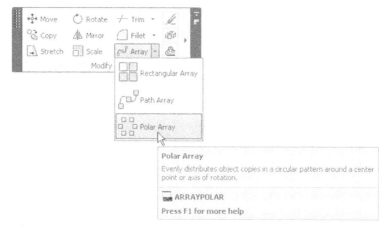

- The following prompts will be shown:

```
Select objects:
Type = Polar  Associative = Yes
Specify center point of array or [Base point/Axis of
rotation]:
Enter number of items or [Angle between/Expression]
<4>:
Specify the angle to fill (+=ccw, -=cw) or
[Expression] <360>:
```

- The first prompt asks you to select the desired objects; press [Enter] when you're done. The second prompt shows the type of array and that associativity is on. The third line asks you to specify the center point of the array and the number of items (you can also use the Angle between items instead; either of these will work with the last prompt). Finally, input the angle to fill (a positive value means CCW). To end the command, press [Enter].
- By default AutoCAD will rotate objects as they are copied around the center point.
- Once the Array command is finished, because it's associative, when you select any object of the array you will see something like the following:

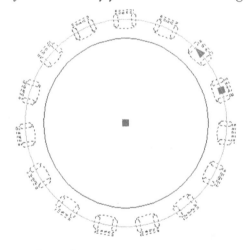

- The first grip at the right:

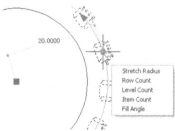

- This grip is for stretching the radius, changing the row count, changing the level count (for 3D only), changing the item count, and finally, changing the fill angle. The following is an example of changing the row count:

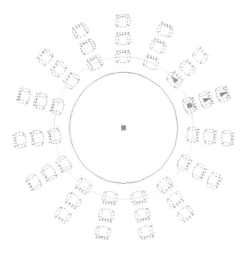

- Two more grips will appear to specify row spacing and row count, or total row spacing.

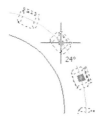

- This grip will allow you to specify the angle between items. The last grip at the center will allow you to move the whole array anywhere you want.
- Meanwhile, a new context tab named Array will appear, which looks like the following:

- This tab will help you modify any of the values by typing them in.
- Levels is for Array in 3D.

ARRAYING OBJECTS USING THE POLAR ARRAY

Practice 4-9

1. Start AutoCAD 2012.
2. Open **Practice 4-9.dwg**.
3. Using the Polar Array create the following shape:

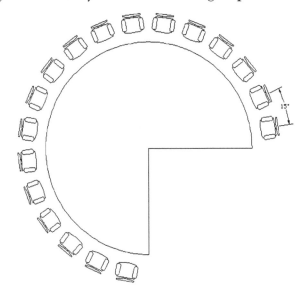

4. Save and close the file.

4.8 LENGTHENING OBJECTS

- Using this command you can add length or subtract length from objects using different methods. To issue this command, go to the **Home** tab, locate the **Modify** panel, and then select the **Lengthen** button:

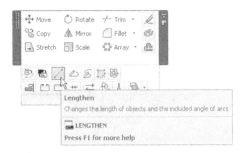

- You will see the following prompt:

```
Select an object or [Delta/Percent/Total/Dynamic]:
```

- If you click an object, AutoCAD will report the current length using the current unit. There are multiple methods to lengthen (or shorten) objects in AutoCAD.
- The first method is called **Delta**, and it allows you to add (remove) to (from) the current length. A positive value means adding, and a negative value means subtracting.
- This is the prompt you will see if you choose the Delta option:

```
Enter delta length or [Angle] <0.0000>:
```

- The second method is called **Percentage**, and it allows you to add (remove) to (from) the length by specifying a percentage of the current length. To add more length input a value more than 100; to remove length input a value less than 100.
- You will see the following prompt if you choose the Percentage option:

```
Enter percentage length <100.0000>:
```

- The third method is **Total**, which allows you to input a new total length of the object. This means the object will add length if the new value is greater than the current value, and vice-versa.
- You will see this prompt if you choose the Total option:

```
Specify total length or [Angle] <1.0000)>:
```

- The fourth method is **Dynamic**, and it allows you to increase/decrease the length dynamically using the mouse. You will see the following prompt:

```
Specify new end point:
```

 - You lengthen/shorten a single object using a single method per command.

4.9 JOINING OBJECTS

- The Join command is a great command because it will help join lines to other lines, arcs to arcs, and polylines to polylines.

- To issue this command, go to the **Home** tab, locate the **Modify** panel, and select the **Join** button:

- AutoCAD will show the following prompts:

```
Select source object:
Select lines to join to source:
Select lines to join to source:
1 line joined to source
```

- The above prompts are for joining lines, and they may differ for arcs and polylines. There are some conditions for the joining to succeed:
  - If you want to join lines they should be always collinear.
  - If you want to join arcs the collective arcs should be part of the same circle.
  - While joining arcs, there a special prompt asking if you want to create a circle (this is the only command that will allow you to create a circle from an arc).
  - You can join polylines to lines and arcs, but all segments should be connected.

LENGTHENING AND JOINING OBJECTS

 Practice 4-10

1. Start AutoCAD 2012.
2. Open **Practice 4-10.dwg**.
3. Using the Lengthen command, make the total length of the line at the upper left = 1.75".

4. Using the Lengthen command, make the arc length at the right side of the shape 200% of the current length.
5. Using the Join command, join the arc at the top and the line at its right.
6. Using the Join command convert the two arcs at the top and at the bottom to a full circle.
7. Using the Join command join all the objects to the polyline.
8. You should get the following shape:

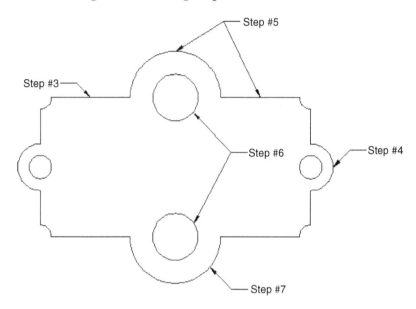

9. Save and close the file.

NOTES:

CHAPTER REVIEW

1. You can fillet using more than one fillet radius using the same command.
   a. True
   b. False
2. There are three types of arrays.
   a. True
   b. False
3. If you hold _____ while filleting you will get radius = 0.0 regardless of the current radius.
4. Which one of the following is not related to the Chamfer command?
   a. Distance 1 and Distance 2
   b. Distance and Angle
   c. Distance and Radius
   d. Trim and No trim
5. While using the Lengthen command, the option to double the current length without knowing it is:
   a. Delta
   b. Total
   c. Percent
   d. Dynamic
6. You can Offset using offset distance and _____.
7. In the Trim command you will select the _____ as a first step, then you will press [Enter].
8. While you are using the Extend command if you hold _____ you will convert the command to Trim.
   a. [Ctrl]
   b. [Ctrl] + [Shift]
   c. [Shift]
   d. [Alt] + [Ctrl]

CHAPTER REVIEW ANSWERS

1. a
3. [Shift]
5. c
7. Cutting Edge(s)

LAYERS AND INQUIRY COMMANDS

Chapter **5**

In This Chapter

◇ What are layers in AutoCAD?
◇ How to create and set layer properties
◇ What are layer controls?
◇ Using the Layer Properties Manager
◇ How to use Quick Properties and Properties
◇ What are the inquiry commands and how do you use them?

5.1 WHAT ARE LAYERS IN AUTOCAD?

- Layers are the most important way to organize and control your AutoCAD drawings. Managing layers means managing the drawing. So, what are layers in AutoCAD? Layers are much like a transparent piece of paper in which you will draw part of the drawing using a certain color, linetype, and lineweight. Each object on a layer will hold the properties of this layer, meaning the object will have the same color, linetype, and lineweight of the layer it resides in. This setting is called **BYLAYER**, which means we control the drawing through the layers rather than the objects.
- Each layer should have a name; this is considered the first step of the layer-creation process. Proper naming should adhere to the following rules:
  - Name length should not exceed 255 characters.
  - You can use any letter (small or capital) in the name.
  - You can use any number in the name.
  - You can use (-) hyphen, (_) underscore, and ($) dollar sign in the name.
- A unique layer exists in all AutoCAD drawings called 0 (zero). This specific layer can't be deleted or renamed, but other layers can be deleted and renamed.

- The layer at the top of the pile is the only layer we can draw on; this layer is called the current layer. So, as a rule of thumb, make the desired layer current first and then start drawing. To start building up your layers, go to the **Home** tab, locate the **Layers** panel, and then click the **Layer Properties** button:

- You will see the following palette, which is called the **Layer Properties Manager**:

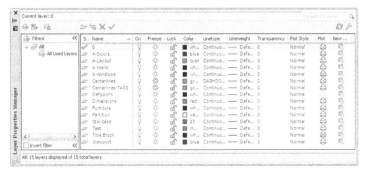

- Palettes in AutoCAD have more features than the normal dialog box:
  - The dialog box has two states: displayed on the screen or closed. The palette can also be displayed but hidden, which keeps you from having to open and close the box:

  - Contrary to most dialog boxes, palettes can be resized horizontally, vertically, and diagonally:

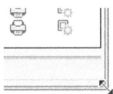

- Dialog boxes can't be docked, but palettes can be docked at the four sides of the screen.

5.2 CREATING AND SETTING LAYER PROPERTIES

- In this part of Chapter 5, we will learn how to do the following things:
  - Create a new layer
  - Set a color for a layer(s)
  - Set the linetype for layer(s)
  - Set a lineweight for layer(s)
  - Set the current layer

5.2.1 How to Create a New Layer

- This command will allow you to add a new layer to the current drawing. Using the **Layer Properties Manager**, click the **New Layer** button:

- This will create a new layer with the temporary name *Layer1*. The **Name** field will be highlighted. Type the desired name of the layer. You should always stick to good naming conventions; i.e., a layer containing doors should be named Doors.
- All the settings discussed as follows require you to select layer(s) in the **Layer Properties Manager**. You select layers in AutoCAD just like any other software running under the Windows OS. You can hold the [Ctrl] key and/or [Shift] to select multiple layers.

5.2.2 How to Set a Color for a Layer(s)

- You can use one of the 256 colors available in AutoCAD. The first seven colors can be set using its name or its number (from color 1 to color 7); they are:
  - Red (1)
  - Yellow (2)

- Green (3)
- Cyan (4)
- Blue (5)
- Magenta (6)
- Black/White (7)
  - Other colors should be set only using their number.
  - To set the color for a layer, take the following steps:
    - Using the **Layer Properties Manager**, select the desired layer(s).
    - Using the **Color** field, click the icon of the color, and you will see the following dialog box:

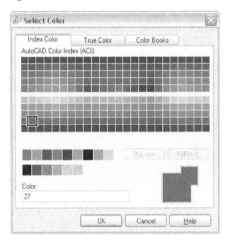

    - Select the desired color (or you can type the name/number in the **Color** field), then click the **OK** button to end this action.
  - Another way to set up (or modify) a layer's color is by using the pop-up list in the **Layers** panel, as shown below:

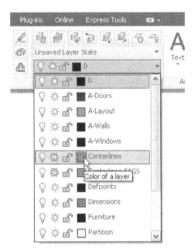

5.2.3 How to Set the Linetype for Layer(s)

- There are two linetype files that come with AutoCAD 2012; they are acad.
  lin, and acadiso.lin. Since these two linetype files are not adequate for all
  types of engineering, designing, and drafting, you should consider buying
  more linetype files available on the market.
- Contrary to colors, linetypes are not loaded with the current file. So, accordingly, you should load the desired linetypes when needed. To set the linetype
  for a layer, take the following steps:
  - Using the **Layer Properties Manager**, select the desired layer(s).
  - Using the **Linetype** field, click the name of the linetype, and you will see
    the following dialog box:

  - If your desired linetype is listed, then select it. If it is not, you need to
    load it. Click the **Load** button, and you will see the following dialog box:

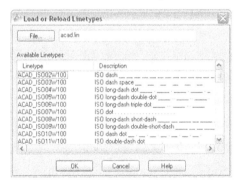

  - Browse for your desired linetype, select it, then click **OK**. Now the
    linetype is loaded, and you can select it and click **OK**.

5.2.4 How to Set a Lineweight for Layer(s)

- This option will allow you to set the lineweight for layer(s). All objects in
  AutoCAD (except polylines with width) have a default lineweight, which is

0 (zero), but you can set the lineweight for objects through their layers. To do so take the following steps:

- Using the **Layer Properties Manager**, select the desired layer(s).
- Under the **Lineweight** field, click the lineweight icon, and the following dialog box will appear:

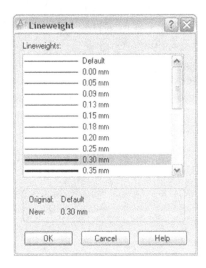

- Select the desired Lineweight, and click **OK.**
- To see the lineweight on the screen use the status bar and click the **Show/ Hide Lineweight** button on:

5.2.5 How to Set the Current Layer

- There are several ways to make a layer the current layer.
- The easiest way is to use the layer pop-up list in the Layers panel, as shown in the following:

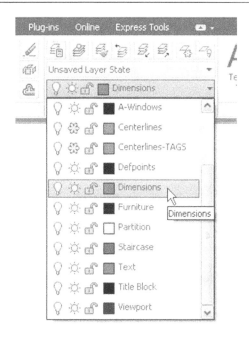

- Another way is to use the **Layer Properties Manager** palette, then double-click the status of the desired layer's name.
- The longest way is to use the **Layer Properties Manager** palette, select the desired layer, and then click the **Set Current** button:

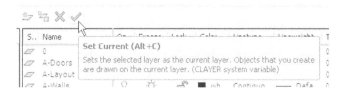

CREATING AND SETTING LAYER PROPERTIES

Practice 5-1

1. Start AutoCAD 2012.
2. Open **Practice 5-1.dwg**.
3. Create a new layer and call it Centerlines; the color should be yellow and the linetype should be Center. Make it current and draw two lines, from 8,14 to 16,14, and the other line from 12,11 to 12,17.

4. Create another layer and call it Hidden; the color should be 9 and the linetype should be Dashed. Make it current and draw two lines, from 3.75, 4.5 to 3.75, 8.5, and the other line from 20.25, 4.5 to 20.25, 8.5.

5. You will have the following shape:

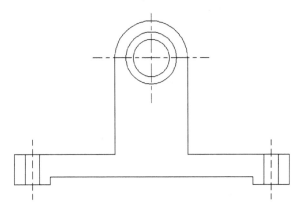

6. Save and close the file.

5.3 LAYER CONTROLS

- The commands discussed here help you have full control over layers, because, after all, layers are the primary tools you use to control your drawings. We will learn how to control the visibility of layers, locking layers, plotting layers, deletion of layers, renaming of layers, etc.

5.3.1 Controlling Layer Visibility, Locking, and Plotting

- AutoCAD provides controls to show/hide layers (Freeze and Off), lock and unlock layers, plot and no plot layers. See the following example:

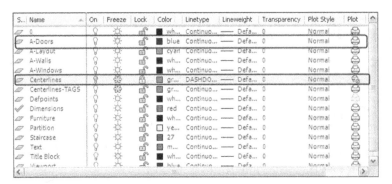

- In the above example, you can see that:
  - The **A-Doors** layer is *On*, *Thawed*, *Unlocked*, and can be *Plotted*.
  - The **Centerlines** layer is *Off*, *Frozen*, *Locked*, and *Not Plotted*.
- When a new layer is created, by default, it will be On, Thawed, Unlocked, and Plotted. You can hide the contents of layers by turning them off or freezing them. Freeze has a deeper effect than off, since objects in a frozen layer will not be considered in the drawing, and the drawing size will be temporarily smaller (you can use Freeze to make the drawing size smaller if the drawing loads slowly).
- If you try to freeze the current layer, AutoCAD will show the following message:

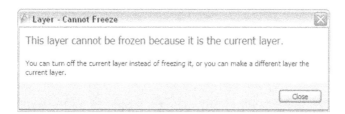

- But if you try to turn off the current layer, you will see the following message:

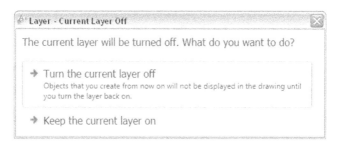

- If you lock a layer, then objects reside in it, and it will not be selected with any modifying commands. Objects in a locked layer will be faded, and when you get close to them you will see a small lock icon appear that tells you this layer is locked. See the following illustration:

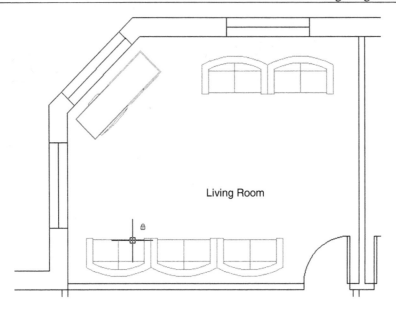

Living Room

- If you choose No Plot for a layer, then objects in this layer will be displayed but not plotted.
- On/Off, Thaw/Freeze, Lock/Unlock can be controlled using the pop-up list in the Layer panel and **Laycr Properties Manager**, while Plot/No Plot can be controlled only in the **Layer Properties Manager**.

5.3.2 Deleting and Renaming Layers

- AutoCAD will not delete a layer that contains objects. AutoCAD will delete only empty layers. In order to delete layers, take the following steps:
  - In the **Layer Properties Manager** palette select the desired layer(s).
  - Press [Del] on the keyboard or click the **Delete Layer** button:

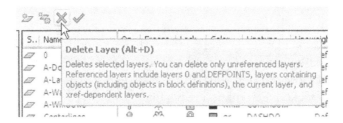

- If you try to delete a layer that contains objects, you will see the following message:

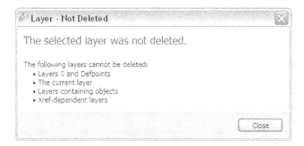

- To rename a layer, take the following steps:
- Select the desired layer.
- Click the name with a single click, and you will see the following:

- The name will become editable; type in a new name and press [Enter].

5.3.3 How to Make an Object's Layer the Current Layer

- This is the fastest way to make a layer the current layer! After issuing the command, select an object that resides in this layer, and follow the steps below:
  - Go to the **Home** tab, locate the **Layers** panel, and then click the **Make Object's Layer Current** button:

- You will see the following prompt:

```
Select object whose layer will become current:
```

- Select the desired object. See what the current layer is. You will find it became the object's layer (even without you knowing the object's layer).

5.3.4 How to Undo Layers Actions Only

- The function to undo the layers actions only is called **Layer Previous**. This function will help you restore previous states of the layers (such as Freeze, Thaw, On, Off, etc.) without affecting other parts of the drawing or other modifying commands. Take the following steps:
  - Change the layer states as needed.
  - Go to the **Home** tab, locate the **Layers** panel, and then click the **Previous** button:

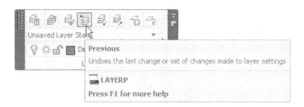

- In the Command window, you will see the following message:

```
Restored previous layer status
```

5.3.5 Moving Objects From One Layer to Another

- All similar objects must reside in the same layer, but mistakes happen. If you draw on the wrong layer, and you want to move the objects to the right layer, you can use the **Match** command. Do the following:
  - Go to the **Home** tab, locate the **Layers** panel, and then click the **Match** button:

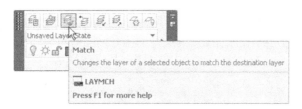

- The following prompts will appear:

```
Select objects to be changed:
Select object on destination layer or [Name]:
```

- This command contains two prompts; using the first you will select the object mistakenly drawn in the wrong layer, then at the second prompt, you will either select the object that resides in the right layer, or simply type its name.

- When you're done you will see something like the following in the Command window:

```
8 objects changed to layer "Dimensions"
```

5.4 USING THE LAYER PROPERTIES MANAGER

- While you are in the **Layer Properties Manager** palette you can do several actions. If you select a layer and then right-click, you will see a menu like the following:

- In this menu you can do any or all of the following:
  - Set the current layer
  - Create a new layer
  - Delete a layer
  - Select All layers
  - Clear the selection
  - Select All but Current
  - Invert the selection (unselect the selected, and vice-versa)
- You also have the option to show or hide the filter tree:
  - Show Filter Tree (by default it is turned on)
  - Show Filters in Layer List (by default it is turned off)

- If you turned off **Show Filter Tree**, you will be allowed to see more information about your layers. See the following:

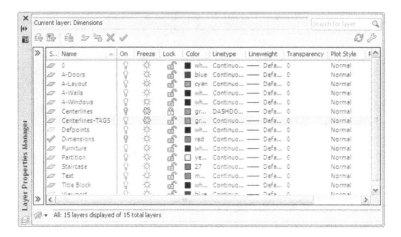

- Another way to do the same thing is by clicking the small arrows at the top-right part of the filter tree pane:

LAYER CONTROLS

 Practice 5-2

1. Start AutoCAD 2012.
2. Open **Practice 5-2.dwg**.
3. Make layer 0 the current layer.
4. Freeze layers Centerlines and Hatch.
5. Lock layer Furniture. What happened to the color?

_____

6. Get closer to any object in the Furniture layer, and notice the icon that appears when you get close.
7. Try to erase one of the objects in layer Furniture. What message do you receive?_____

8. Using the **Layer Properties Manager**, try to freeze layer 0 (the current layer) What message do you receive? _____
9. Try to rename layer 0. What message do you receive? _____
10. Rename layer Partition to Inside Wall.
11. Using the Match command, select the two doors at the bottom (their colors are black and white) to match one of the blue doors.
12. Using the **Layer Properties Manager**, select all the layers, then change their color to black or white), then close the **Layer Properties Manager**.
13. Click Layer Previous; what happened? _____
14. Click the Make Object's Layer Current button, and click one of the yellow lines. What is the current layer now? _____
15. Save and close the file.

5.5 CHANGING AN OBJECT'S LAYER, QUICK PROPERTIES, AND PROPERTIES

- In this part of Chapter 5, we will learn how to control an object's properties.

5.5.1 Reading Instantaneous Information About an Object

- AutoCAD provides instant information about any object when you hover your mouse over it. See the following illustration:

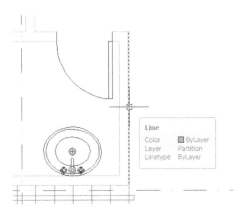

- You can see that AutoCAD is showing the type of object (Line), its color, the layer it's in, and its linetype.

5.5.2 How to Move an Object From One Layer to Another Layer

- All objects in AutoCAD should reside in a layer. To move an object from one layer to another, take the following steps:
  - Click the desired object(s).
  - Go to the **Home** tab, locate the **Layers** panel, and find the layer name in the pop-up list. Sometimes this will be blank; this happens when your desired objects reside in different layers. Click the layer's pop-up list, and select the new layer name.
  - Press [Esc] once to deselect all the selected objects.

5.5.3 What is Quick Properties?

- This is the first of two commands that allow you to change the properties of selected objects. An object's properties differ depending on the object type, and the line properties are different depending on if it's an arc, circle, or polyline.
- **Quick Properties** will pop up on the screen whenever you select objects without issuing a command (using grips). You can turn it off using the status bar, as shown below:

- When you select object(s) you will see the following:

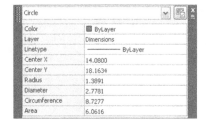

- You will see information including:
  - Color
  - Layer
  - Linetype
  - Coordinates of Center Point (X & Y)
  - Radius and Diameter
  - Circumference and Area

- If you select more than one object of the same type, you will see this:

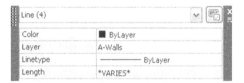

- But if you select more than one object of different types, you will see this:

- You will see All (number), which means you are seeing eight objects (in our example) from different types, but you can see a breakdown of the selected objects by clicking the pop-up list, as shown below:

- When you select a single object type, you can change the general and specific object's properties.

5.5.4 What is Properties?

- As the name indicates, Quick Properties is a fast way to change the properties of object(s). But the **Properties** command is much more comprehensive; all the rules we discussed concerning Quick Properties are applicable here as well.
- To issue the **Properties** command, select the desired object(s), right-click, and then select **Properties**.
- You can also double-click any object to get the Properties palette. Accessing Properties this way has two drawbacks; first, it will work for a single object only, and second, some objects such as polylines, blocks, hatches, and text will interpret this action as an edit. Either way, you will see something like the following:

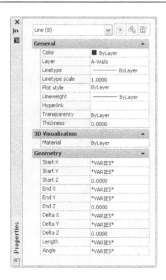

- As you can see, there is much more information displayed here than in Quick Properties, which gives you the chance to make more changes to your selected object(s).
- Some of the information is shaded, which means you cannot change it.

CHANGING AN OBJECT'S LAYER AND USING QUICK PROPERTIES AND PROPERTIES

Practice 5-3

1. Start AutoCAD 2012.
2. Open **Practice 5-3.dwg**.
3. Hover over one of the circles of the centerlines. What is the name of the layer? _____
4. Select all the circles and text inside them and move them to layer Centerlines.
5. Change the properties of the circles to Continuous.
6. Delete layer Centerlines-TAGS.
7. Zoom to the lower door; you will find two red lines at the right and at the left. Move them from layer Dimensions to layer A-Walls.
8. Save and close the file.

5.6 AN INTRODUCTION TO INQUIRY COMMANDS

- The main purpose of this set of commands is to allow you to measure the length between two points, find the radius of a circle or arc, and measure the angle, area, or volume of 3D objects.
- You will use this set of commands to make sure that your drawing is correct and according to the design intent.
- To issue these functions go to the **Home** tab and locate the **Utilities** panel.

5.7 MEASURING DISTANCE

- This command will allow you to measure the distance between two selected points. To issue this command, go to the **Home** tab, locate the **Utilities** panel, and click the **Distance** button:

- You will see the following prompts:

```
Specify first point:
Specify second point:
```

- Select the desired points, and AutoCAD will display something like the following:

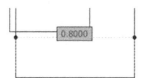

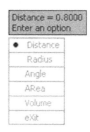

- You will also see the following result in the Command window:

```
Distance = 0.80, Angle in XY Plane = 0, Angle from XY
Plane = 0
Delta X = 0.80, Delta Y = 0.0000, Delta Z = 0.0000
```

5.8 CHECKING THE RADIUS

- This command will allow you to check the radius (and parameter as well) of a drawn circle or arc. To issue this command, go to the **Home** tab, locate the **Utilities** panel, and select the **Radius** button:

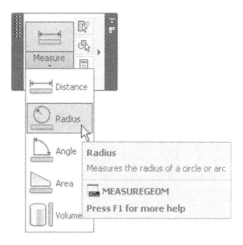

- AutoCAD will display the following prompt:

```
Select arc or circle:
```

- Select the desired arc or circle, and you will see the following:

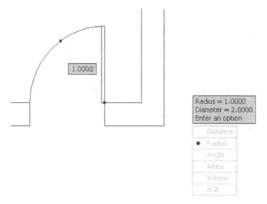

- You will also see the following in the Command window:

```
Radius = 1.000
Diameter = 2.000
```

5.9 MEASURING THE ANGLE

- This command will allow you to measure the angle (between two lines, the included angle of an arc, or two points and the center of the circle). To issue this command go to the **Home** tab, locate the **Utilities** panel, then select the **Angle** button:

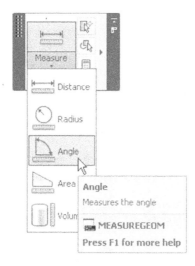

- You will also see the following prompt:

```
Select arc, circle, line, or <Specify vertex>:
```

- Select the desired objects (whether two lines, an arc, or points on a circle), and AutoCAD will display the following:

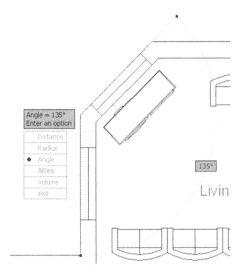

- You will also see something like the following in the Command window:

```
Angle = 135°
```

5.10 MEASURING AREA

- This command will allow you to measure areas, whether simple (areas that have no islands inside) or complex areas (areas that have islands inside). AutoCAD can measure areas between points (assuming lines and arcs connecting them) or objects (such as circles, closed polylines, etc.). To issue this command go to the **Home** tab, locate the **Utilities** panel, and then select the **Area** button:

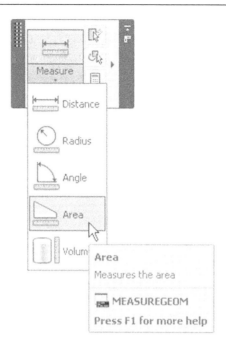

- You will also see the following prompt:

```
Specify first corner point or [Object/Add area/
Subtract area/exit] <Object>:
```

5.10.1 How to Calculate Simple Area

- The definition of simple area is any closed area without any objects (islands) inside it. AutoCAD will assume you want to measure a simple area if you start by specifying points or selecting objects. If you start specifying points, AutoCAD will assume there are either lines or arcs connecting them. For lines you will see the following prompts:

```
Specify next point or [Arc/Length/Undo]:
Specify next point or [Arc/Length/Undo]:
Specify next point or [Arc/Length/Undo/Total] <Total>:
Specify next point or [Arc/Length/Undo/Total] <Total>:
```

- If you see an arc in your area, simply change the mode to Arc, and you will see prompts identical to the Polyline command. Keep specifying points of lines or arcs, until you press [Enter]. You will see the Total value of the measured area in the Command window and the following:

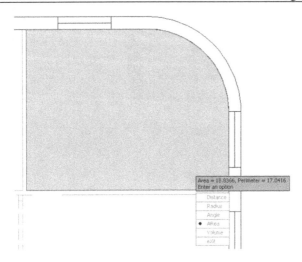

- You will also see the following displayed in the Command window:

```
Area = 18.8366, Perimeter = 17.0416
```

- If you have a simple area and the parameter is a single object, you can select the object rather than specifying points. Either right-click and select the **Object** option, or type **o** in the Command window. You will see the following prompt:

```
Select objects:
```

- Select the object you want to measure, and you will see the following:

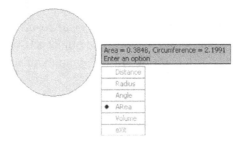

- You will also see the following in the Command window:

```
Area = 0.3848, Circumference = 2.1991
```

5.10.2 How to Calculate Complex Area

- The definition of complex area is any closed area with objects (islands) inside it. To tell AutoCAD you want to calculate a complex area, you <u>have</u> to start with either **Add area** or **Subtract area**.
- If you start with Add area or Subtract area, AutoCAD will start with area = 0, which will allow you to add areas, then subtract areas as needed. For Add area mode you will see the following prompt:

```
Specify first corner point or [Object/Subtract
area/exit]:
```

- Specify the area using the same methods discussed above (points or object), then switch to Subtract area mode, and you will see the following prompt:

```
Specify first corner point or [Object/Add area/eXit]:
```

- AutoCAD will give you a sub-total after adding or subtracting each area. When you are done, press [Enter] twice to end the command, and you will get the final net area.
- You will see something like the following:

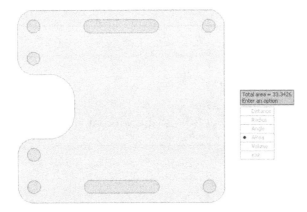

INQUIRY COMMANDS

  Practice 5-4

1. Start AutoCAD 2012.
2. Open **Practice 5-4.dwg**.

3. Freeze all layers except A-Walls (make layer A-Walls current, then select all layers except the current, and freeze them).
4. Measure the length of the slanted wall from inside, and enter the information given by AutoCAD here:
   a. Length = _____ (1.6971)
   b. Angle in XY plane = _____ (45 or 135)
   c. Delta X = _____ (1.2000)
   d. Dleta Y = _____ (1.2000)
5. Thaw layer Partition.
6. Measure the horizontal and vertical lengths of the room at the upper-right (you can use the Nearest and Perpendicular OSNAPS), and enter them here:
   a. Horizontal Distance = _____ (5.05)
   b. Vertical Distance = _____ (3.9)
7. Measure the inside area of the room at the lower-right part of the plan, and enter the measurements here:
   a. Area = _____ (21.8016)
   b. Parameter = _____ (18.2416)
8. Save and close the file.

INQUIRY COMMANDS

 Practice 5-5

1. Start AutoCAD 2012.
2. Open **Practice 5-5.dwg**.
3. Calculate the net area of the shape without all the inside objects, and enter it here: Area = _____ (33.3426).
4. Save and close the file.

NOTES:

CHAPTER REVIEW

1. Which one of the following is not true about layer names?
   a. They should not exceed 256 characters
   b. Spaces are allowed
   c. $ is allowed
   d. $ is not allowed
2. You can't _____ the current layer.
3. You can _____ the current layer.
4. You can undo layer actions only.
   a. True
   b. False
5. The Area command can calculate only an area without any islands inside it.
   a. True
   b. False
6. The following are facts about layer 0 (zero) *except*:
   a. You can't rename it
   b. You can't set a new color for it
   c. You can't delete it
   d. It is in all AutoCAD files
7. Delta X is one of the properties available in the _____ command.
8. Most objects will respond to _____ to display the Properties palette.

CHAPTER REVIEW ANSWERS

1. d
3. Turn off
5. b
7. Measure distance

6 BLOCKS AND HATCHES

Chapter

In This Chapter
◇ Blocks and how to define them
◇ How to use (insert) blocks
◇ How to explode and convert blocks
◇ Design Center and Tool Palettes
◇ How to edit blocks
◇ How to hatch in AutoCAD
◇ How to control hatches in AutoCAD
◇ How to edit hatches

6.1 WHAT ARE BLOCKS?

- In your daily work there will often be a shape you need repeatedly, and you have two ways to get it. You can draw it each time you need it, or you can draw it once, and save it as a block (the block will be a single object) and you can use (insert) it as many times as you wish in the current file and other files as well.
- There are many benefits to using blocks, including:
  - The file size will be smaller because each block will be counted as a single object.
  - Standardization within the company.
  - Speed of completing drawings.
  - Design Center and Tool Palettes.

6.2 HOW TO CREATE A BLOCK

- Use the following steps to create a block:
    - Draft the shape that you want to create a block from in layer 0 (zero). Layer 0 will enable the block to inherit the properties (color, linetype, lineweight) of the layer it resides in.
    - Make sure to control the "Block unit," which enables AutoCAD to automatically scale the block to appear as the right size in any other drawing.
    - Draft the shape you want to create a block from in its real-world dimensions.
- Let's assume we draw the following shape:

- Now you are ready to issue the command. Go to the **Insert** tab, locate the **Block Definition** panel, and select the **Create Block** button:

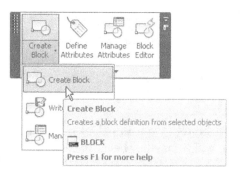

- You will see the following dialog box:

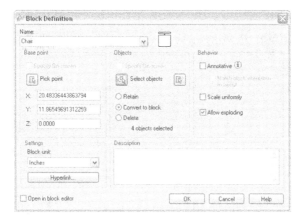

- ■ Do the following:
  - • Type a name for the block; it should not exceed 255 characters (use only numbers, letters, -, _, $, and spaces).
  - • Specify the **Base point**, either by typing X, Y, Z coordinates, or click the **Pick point** button to input the base point graphically.
  - • Click the **Select objects** button to select the desired objects.
  - • From the drawn object, AutoCAD will create the needed block, but what do you do with the object afterwards? You should select one of the three choices available, and either Retain (leave) the objects as they are, Convert them to a block, or Delete them:

  - • Select whether the block will be Annotative (a feature we will discuss in Chapter 9), whether to scale uniformly in both X and Y, and whether to allow the block to be exploded:

- Select the **Block unit**. This will tell AutoCAD what each AutoCAD unit used in this block will equal to. This will help AutoCAD in the Automatic Scaling feature:

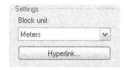

- Enter a block description.
- Turn the checkbox "Open in a block editor" off because this is an advanced feature that is used for creating dynamic blocks.
- Once you are finished inputting all the above data, click **OK.**
- Later, you will use (insert) the block, but this will only be a copy of the block. The original block definition will stay intact.

6.3 HOW TO USE (INSERT) BLOCKS

- After creating the block, you are ready to use (insert) the block in the current drawing. Make sure you are in the right layer, and the drawing is ready to accept the block (make sure the door openings are created before inserting the door, for example).
- Now you are ready to issue the command. Go to the **Insert** tab, locate the **Block** panel, and then click the **Insert** button:

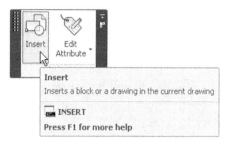

- You will see the following dialog box:

- If the block is created in the current file, click the pop-up list to select it. You can also click the **Browse** button to select a file and insert it in the current file as a block.
- You should specify the **Insertion point**, the **Scale**, and the **Rotation** angle either by using the **Specify On-screen** checkbox, or by typing the needed value.
- While using **Scale** you can insert mirror images of the block by using negative values. See the following illustration:

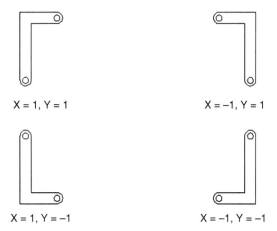

- While you are using **Rotation angle**, remember CCW is always positive.

6.3.1 Automatic Scaling

- When we created the block we input the Block unit, so each unit we use in this block will represent a certain unit. In order for this circle to be complete you have to control the drawing file unit. Go to the **Application Menu**, select **Drawing Utilities**, and then select **Units**. You will see the following dialog box:

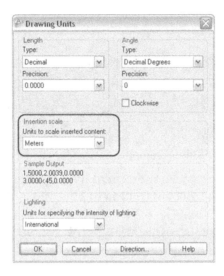

- Under **Units to scale inserted content** select the desired unit that will represent your drawing file unit. AutoCAD will convert the block unit to the drawing unit and show it in the **Insert** dialog box under **Block unit**.

6.3.2 Block Insertion Point OSNAP

- After you insert a block, click it to see the following:

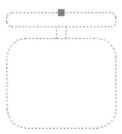

- The whole block is one unit, and only one grip is highlighted, which is the insertion point. There is a specific OSNAP to snap to this point called **Insertion** (or insert depending on where you are looking). See the following illustration:

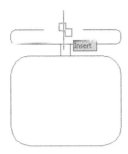

CREATING AND INSERTING BLOCKS

 Practice 6-1

1. Start AutoCAD 2012.
2. Open **Practice 6-1.dwg**.
3. There are three shapes at the left drawn in layer 0. From these shapes, create three blocks, and name them Window, Single Door, and Double Door, making the block unit for the three blocks Meter.
4. Using Application Menu/Drawing Utilities/Units, change Units to scale inserted content to Meter.
5. Use the Insert command to insert the three blocks in the proper places using the proper layers, just like the following:

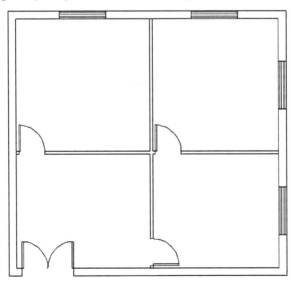

6. Save and close the file.

6.4 EXPLODING BLOCKS AND CONVERTING THEM TO FILES

6.4.1 Exploding Blocks

- You can explode blocks the same way you can explode polylines to lines and arcs. The Explode command will bring them back to their original objects. This practice is not recommended, however, since it's better to keep blocks as one object.
- To issue the command, go to the **Home** tab, locate the **Modify** panel, and then select the **Explode** button:

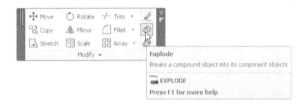

- You will see the following prompt:

```
Select objects:
```

- Select the desired block(s) and press [Enter] to end the command.

6.4.2 Converting Blocks to Files

- In earlier versions of AutoCAD, we used to convert all of our blocks to files in order to use them in other files. This practice was eclipsed by the emergence of Design Center and Tool Palettes in more recent versions.
- To issue the command, go to the **Insert** tab and locate the **Block Definition**, then click the **Write Block** button:

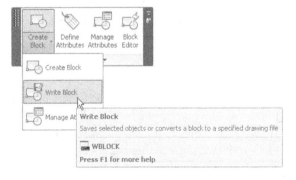

- You will see the following dialog box:

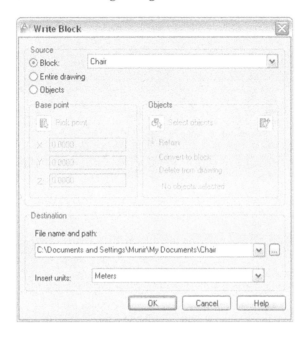

- Select the **Block** option under **Source**. Select the name of the block. Under **Destination** input the file name and path and specify the insert unit. You can use the same dialog box to create a file from the entire drawing or from some of the objects in the current drawing.

6.5 DESIGN CENTER

- Before AutoCAD 2000, there was no direct method for sharing blocks, layers, and other things. In AutoCAD 2000, **Design Center** was introduced as the ultimate solution to this problem. Using Design Center, you can share blocks, layers, dimension styles, text styles, table styles, etc.
- To issue the command go to the **Insert** tab, locate the **Content** panel, and then click the **Design Center** button:

- You will see the following palette:

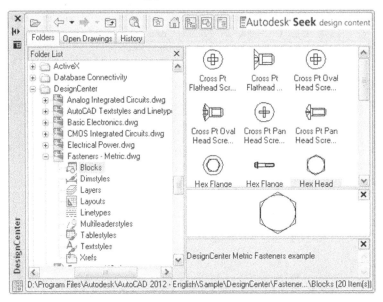

- The left pane of the Design Center is like My Computer in Windows, and it contains all of your drives, folders, and files. This will enable you to locate the file that contains the needed blocks, layers, etc. When you locate the file, click the plus sign at the left of the file name and a list will come up. See the following illustration:

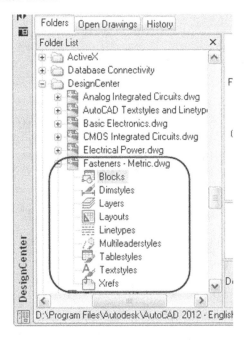

- Select the desired content. If you select Blocks, in the right pane you will see the shapes and the names. There are three methods to copy blocks to your current drawing using Design Center:
  - Drag-and-drop using the left mouse button to insert the block in the current file.
  - Drag-and-drop using the right mouse button. When you release it, you will see the following menu. Select the **Insert Block** option to make the Insert dialog box pop up:

  - Double-click and the Insert dialog box will open.

EXPLODING, CONVERTING, AND USING DESIGN CENTER

 Practice 6-2

1. Start AutoCAD 2012.
2. Open **Practice 6-2.dwg**.
3. Make layer Furniture the current layer.
4. Start Design Center.
5. Locate your AutoCAD 2011 folder, then go to the following path: \Sample\ Design Center.
6. Locate the file Home-Space Planner.dwg, and choose Blocks to insert the furniture shown below.
7. Make layer Toilet the current layer.
8. Locate the file House Designer.dwg, and choose Blocks to insert the toilet furniture as shown below:

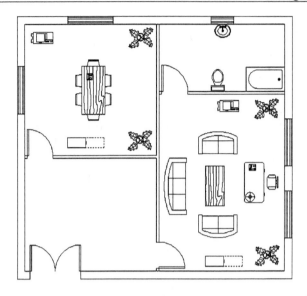

9. Start the WBlock command, and select the Single Door block to convert it to a file, and save it in your practice folder.
10. Save and close the file.

6.6 TOOL PALETTES

- Using Design Center to share data between files is a huge help, but there are still some lingering issues. You have to remember the current layer, the scale, the rotation angle, and search for the desired file each time you need to copy something from it. In 2004, AutoCAD introduced to Tool Palettes, which take blocks to the next level.
- Tool Palettes allow you to store any type of object and then retrieve it in any opened file. You can also control the object's properties, so next time you drag it into your file, you will not have to worry about layers, scale, or rotation angles. You can keep several copies of the same block, and each can hold different properties.
- To start the command, go to the **View** tab, locate the **Palettes** panel, and then click the **Tool Palettes** button:

- You will see something like the following:

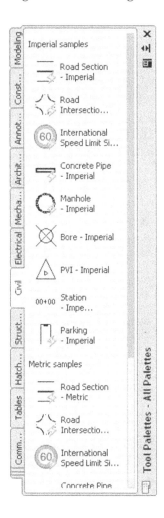

- AutoCAD includes a few tool palettes, but you can also create your own.

6.6.1 How to Create a Tool Palette from Scratch

- This method will create an empty palette, and you can fill it using different methods. To do this right-click over the name of any existing tool palette, and you will see the following menu:

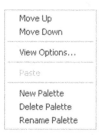

- Choose the **New Palette** option, and a new empty palette will be created, allowing you to name it. Type in the name of the new palette as shown below:

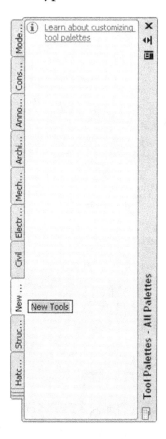

6.6.2 How to Fill the New Palette with Content

- You can fill the new empty palette with content using the drag-and-drop method from different drawings to the palette. For example, say you opened one of your colleague's drawing files and you discovered that she created several new blocks. Simply click the block, avoiding the grips, hold, and drag it to the palette. You can do the same thing with normal objects (lines, arcs, circle, polylines, hatches, tables, dimensions, etc.).

6.6.3 How to Create a Palette from Design Center Blocks

- AutoCAD allows you to create a palette from all blocks in a file using Design Center. In order for this to work, make sure that both Design Center and Tool Palettes are displayed in front of you. Go to your desired file in the left pane of the Design Center, right-click, and you will see something like the following:

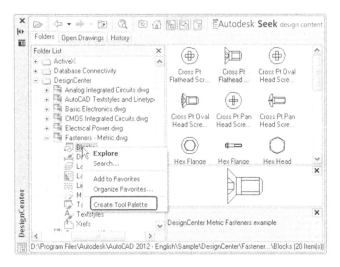

- Select the **Create Tool Palette** option, and a new tool palette with the same name as the file will be created containing all the blocks.
- You can also drag-and-drop any block in any file from the Design Center to any tool palette.

6.6.4 How to Customize Tools Properties

- You can create several copies of blocks and hatches in Tool Palettes using the normal Copy/Paste procedure. Once you have several copies of the same block/hatch, you can change the properties of these copies according to your needs. Follow these steps:

- Right-click on the copied block/hatch, and you will see the following menu:

- Choose the **Properties** option, and you will see something like the following dialog box:

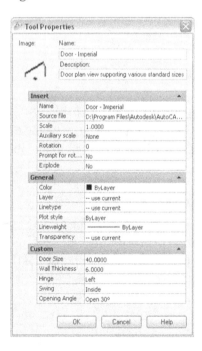

- Type in a new **Name** and **Description**. The properties of the block/hatch are divided into two categories:
  - **Insert** type of properties.
  - **General** type of properties.
- By default, the **General** properties are all **use current**. For a block or a hatch you can specify that whenever you drag from the tool palette the block or hatch will reside in a certain layer (regardless of the current layer) and will have a certain color, linetype, lineweight, etc.
- With this feature, you can create your own tool palettes, which hold all the needed blocks and hatches, customized according to company standards, and designing will become a simple drag-and-drop process. If we know that 30–40% of drawings are blocks and 10–20% are hatches, if you use tool palettes effectively, your drawing time will be reduced significantly.

USING AND CUSTOMIZING TOOL PALETTES

 Practice 6-3

1. Start AutoCAD 2012.
2. Open **Practice 6-3.dwg**.
3. Open Design Center.
4. Locate /Sample/Design Center/Fasteners-US.dwg.
5. Locate the block underneath it, right-click, and then select the Create Tool Palette option.
6. If Tool Palettes is not displayed, it will be displayed with a new palette called Fasteners-US.
7. Close Design Center.
8. Make sure that the current layer is 0.
9. Using the newly created tool palette, locate Hex Bolt ½ in. -side, right-click, and select Properties.
10. Change the layer to Bolts.
11. Create another copy of it, and make the Rotation angle = 270.
12. Using drag-and-drop, drag the two blocks to the proper places as shown below.
13. Using the newly created tool palette, locate Square nut ½ in. -top, and make a copy.
14. Right-click the copy, select Properties, and change the layer to Nut and the scale to 1.5.
15. Using drag-and-drop, drag the new block to the proper places as shown below.

16. Erase the lines to get the following shape:

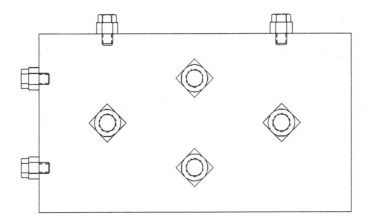

17. Delete the tool palette you created.
18. Close Tool Palettes.
19. Save and close the file.

6.7 EDITING BLOCKS

- We can edit the original block in Block Editor, which is used to create dynamic features of a block (this is an advanced feature of AutoCAD). Block Editor will allow you to redefine the block by adding/removing/modifying the existing objects.
- To issue the command, go to the **Insert** tab, locate the **Block Definition** panel, and then click the **Block Editor** button:

- You will see the following dialog box:

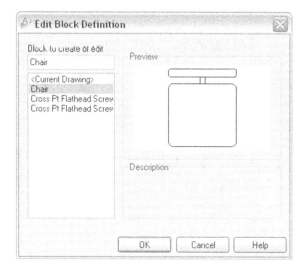

- Select the name of the desired block to edit and then click **OK** to start editing. Another easy way is to double-click one of the blocks.
- When the Block Editor opens, you will see several things take place: the background color will be different, a new context tab called "Block Editor" will appear, and several new panels will appear as well. Ignore all of this, and start adding, removing, or modifying the objects of your block. Once you are done, click the **Save Block** button on the **Open/Save** panel (you can find it at the left), and then click the **Close Block Editor** button in the **Close** panel to end the command.

EDITING BLOCKS

 Practice 6-4

1. Start AutoCAD 2012.
2. Open **Practice 6-4.dwg**.
3. Double-click one of the Single Door blocks.
4. Change the arc linetype to Dashed2.
5. Save the block with the new changes.
6. What happened to the other Single Door blocks?

7. You should have the following shape:

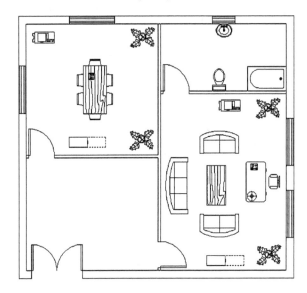

8. Save and close the file.

6.8 HATCHING IN AUTOCAD

- AutoCAD can hatch closed areas and non-closed areas (with the maximum distance defined by the user).
- There are two hatch pattern files that come with AutoCAD: *acad.pat* and *acadiso.pat* (as you can see the hatch pattern file's extension is *.pat).
- There are four pattern types:
  - Solid (single pattern covers the area with a single solid color).
  - Gradient (two gradient colors mixed together in several ways).
  - Pattern (several pre-defined patterns).
  - User defined (the simplest pattern, parallel lines).

6.9 HATCH COMMAND: FIRST STEP

- This command will allow you to put a hatch in your drawing and control all of its properties. The preview is instant. To issue the Hatch command, go to the **Home** tab, locate the **Draw** panel, and then click the **Hatch** button:

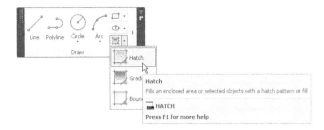

- You will see a new context tab called **Hatch Creation**. You will also see several panels (which will be discussed). Your should locate the **Properties** panel, at the top-left, and select the **Hatch Type** as shown below:

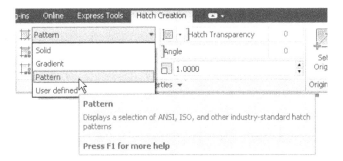

- Once you select the Hatch Type, locate the **Pattern** panel, and AutoCAD will take you to the first pattern in the selected type. For instance, if you select **Gradient** for the Hatch Type, the first pattern in the Pattern panel will be GR_LINEAR, which is the first pattern in the gradient patterns:

- Now, simply go (without clicking) to the desired area to be hatched, and you will see the area filled. At this moment you have two choices:
  - If you like the result, click to choose the area, then go to the **Close** panel, and click **Close Hatch Creation**.
  - If you don't like the result, you can change the properties of the hatch for a different result.

6.10 CONTROLLING HATCH PROPERTIES

- If you clicked inside the area and you don't like the result, you need to alter the properties of the hatch. All of these properties exist in the **Properties** panel; they include:
  - **Hatch Color**: Specify the color of the hatch, or leave it as "Use Current":

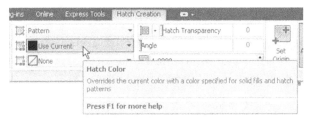

  - **Background Color**: Specify the color of the background, or leave it as "None":

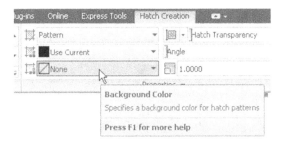

  - **Transparency**: By default, all colors will be with their normal colors, but you can increase the value of transparency (maximum 90) to decrease the intensity of the color (hatch color and background color):

  - **Angle**: Specify the angle of the hatch pattern (there will be no effect with Solid hatching):

- **Scale** (if you chose a user-defined hatch, this will be called Spacing): Input the scale or spacing for the selected hatch pattern:

- **Hatch Layer Override**: By default, the hatch will reside in the current layer. Using this function you can specify the layer you want the hatch to reside in, regardless of the current layer:

- **Double**: This option is only valid if the hatch type is user defined. It controls whether the lines are in one direction or crosshatched:

HATCHING AND CONTROLLING HATCH PROPERTIES

Practice 6-5

1. Start AutoCAD 2012.
2. Open **Practice 6-5.dwg**.

3. Make layer Hatch the current layer.
4. Start the Hatch command.
5. Make sure the following settings are correct:
   a. At the Properties panel, the Hatch Type is Pattern.
   b. At the Pattern panel, ANSI31.
6. Hover over any part of the drawing, and you will see a preview of the hatch. You will notice that the Scale is a little small, so increase it to 2.
7. Choose Hatch background = Yellow.
8. Then select the following areas to hatch:

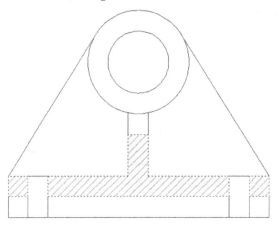

9. Click the Close Hatch Creation button at the right.
10. Start the Hatch command again, and notice that all the options included with the previous hatch are still valid. Change the angle to 90°, then press [Enter] to end the command.
11. You should have the following result:

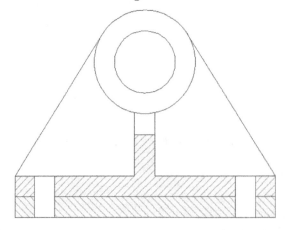

12. Save and close the file.

6.11 SPECIFYING HATCH ORIGIN

- If you want to hatch an area and start the pattern from a certain point and not use the default settings of AutoCAD, then you need to manually set the Hatch Origin. By default, AutoCAD uses 0,0 as the starting point for any hatch, which means you will never know for sure if your hatch will be displayed correctly or not. In order to control the Hatch Origin, make sure you are still in the **Hatch Creation** context tab and locate the **Origin** panel. You will see the following:

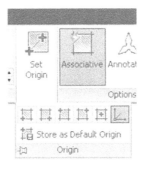

- The first button is **Set Origin**, which will ask you to:

```
Specify origin point:
```

- Specify the desired point. Or you can use the other pre-defined points (lower-left corner, upper-left corner, etc.). You also have the option to save the point you chose this time, for future use, instead of using 0,0.
- See the following example:

Default Orgin   Orgin is the lower-left corner

6.12 CONTROLLING HATCH OPTIONS

- These options will control the outcome of the hatching process. Using these options you will be able to hatch an open area (Gap Tolerance), create separate hatches, and more:

6.12.1 How to Create Associative Hatching

- In AutoCAD, hatching is associative, which means the hatch understands the boundary that it fills! When this boundary changes, the hatch will respond appropriately.
- See the illustration below:

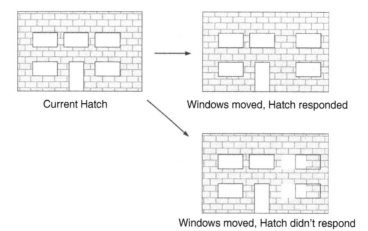

Current Hatch   Windows moved, Hatch responded

Windows moved, Hatch didn't respond

6.12.2 How to Make Your Hatch Annotative

- In Chapter 9 we will discuss printing in AutoCAD, and we will discuss Layouts among other things. One of the features discussed is the **Annotative** feature, and whether you should put hatches, text, and dimensions in Model space or Paper space (layout).

6.12.3 Using Match Properties to Create Identical Hatches

- This option will create an identical hatch from an existing hatch (it will reside in the same layer, and it will also have the same angle, scale, transparency, etc.).
- This option has two buttons associated with it: Use current origin or Use source hatch origin. Both options are self-explanatory:

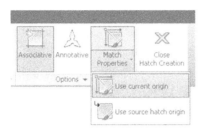

- Select the **Match Properties** button, and you will see the following prompt:

```
Select hatch object:
```

- Select the hatch object you desire to mimic, and you will see the following prompt:

```
Pick internal point or [Select objects/seTtings]:
```

- Click inside the desired area. Keep selecting the area, and when done, press [Enter] to end the command.

6.12.4 Hatching an Open Area

- By default, AutoCAD will hatch only closed areas. But you can ask AutoCAD to hatch an area with an opening. To tell AutoCAD to allow hatching an open area, simply set the **Gap Tolerance,** which will be considered the maximum allowable opening. Any area with an opening bigger than this value will not be hatched:

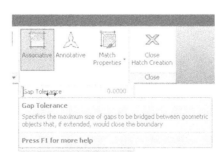

- When you hatch this area, the preview will not be displayed, and you will need to click inside the opened area. You will see the following warning message:

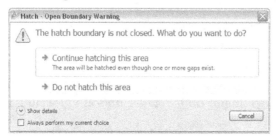

6.12.5 Creating Separate Hatches in the Same Command

- If you are using the same command and you hatched several separate areas, they will be considered a single hatch (single object).
- You can override this default setting by telling AutoCAD that you want separate hatches for separate areas. Simply click this button on:

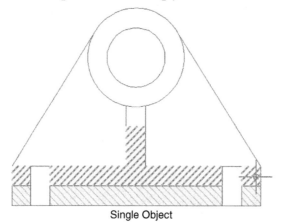

Single Object

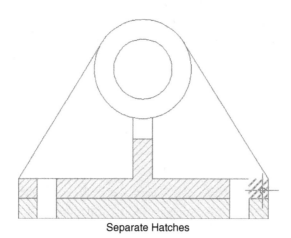

Separate Hatches

6.12.6 Island Detection

■ When you are hatching an area containing several areas (islands) inside it, these inside areas may contain more islands. We want to know how AutoCAD will treat these islands.

■ There are four different choices:

• **Normal Island Detection:** AutoCAD will hatch the first area (the outer one), then leave the second one, hatching the third, and so on.
• **Outer Island Detection:** AutoCAD will hatch the outer area only.
• **Ignore Island Detection:** AutoCAD will ignore all of the inner islands, and hatch the outer area, as if there are no areas inside.
• **No Islands Detection:** This option will turn off the Island Detection feature, which will produce the same result as the Ignore Island Detection option.

6.12.7 Set Hatch Draw Order

■ You can set the draw order of hatches, relative to the other objects, just like any other object in AutoCAD. You have five choices to choose from:

- These are:
  - Do Not Assign (use the default).
  - Send to Back.
  - Bring to Front.
  - Send Behind Boundary.
  - Bring in Front of Boundary.
- See the following example:

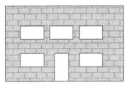

Send Behind Boundary

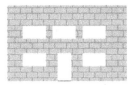

Bring in Front of Boundary

HATCH ORIGIN AND OPTIONS

 Practice 6-6

1. Start AutoCAD 2012.
2. Open **Practice 6-6.dwg**.
3. Start the Hatch command.

4. Using the Options panel, click the Match Properties button, select the hatch of the Toilet, and apply it to the Kitchen; press [Enter] to end the command.
5. Start the Hatch command again, set the Background Color = 40, Scale = 4, Transparency = 0, and hover over Study room. We need to change the Origin point to the lower-left corner of the room. Click inside Study and end the command.
6. Zoom to the upper-right corner of Living Room, and you will see the area is opened. Start the Hatch command again and set the Gap Tolerance = 0.3 (the opening in this drawing is 0.2, so 0.3 is enough), then set to create separate hatches, and click inside Living Room and you will see a warning message. Select the Continue hatching this area option, then click inside Sitting Room and end the command.
7. Thaw layer A-Doors, and notice the two doors of the Living Room and Study; they are not shown properly. To solve this problem, select the hatch, right-click, select the Draw Order option, and select the Send to Back option.
8. Start the Hatch command for the fourth time. For Hatch Type, select Solid, and hatch the outside walls.
9. Zoom to any window of the Kitchen, and move it a short distance. What happened to the hatch? _____
    Why? _____
10. Save and close the file.

6.13 HATCH BOUNDARIES

- If we were discussing older versions of AutoCAD, this panel (the Boundary panel) would be the first panel to discuss, but because of the instantaneous display of the hatch once you are inside the hatch area, this panel is not as important.
- Depending on what you are doing, creating a new hatch or editing an existing one, some of the buttons will be turned off and some of them will be on:

- The **Pick Points** button is always on, which will allow you to pick the areas to hatch, while **Select** and **Remove** will allow you to add/remove more objects to be included in the hatch boundaries.
- If you are editing a hatch (by clicking it), the **Recreate** button will be on. This button will help you recreate the boundary if (for any reason) the boundary was deleted. Simply click the hatch without its boundary, then click the **Recreate** button, and you will see the following prompts:

```
Enter type of boundary object [Region/Polyline]
<Polyline>:
Reassociate hatch with new boundary? [Yes/No] <N>:
```

- The first prompt will ask you to select the desired boundary type, then asks you to re-associate the boundary with the hatch.
- If you select any hatch, AutoCAD will enable you to highlight (**Display**) **Boundary Objects**, so you can edit the boundary. See the following illustration:

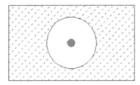 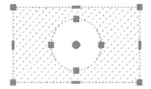

Display Boundary objects = Off   Display Boundary objects = On

- When you create a hatch, AutoCAD normally creates a polyline (or region) that fits the boundary exactly. Once the command ends, AutoCAD will delete it. Using the **Retain Boundary** pop-up list you can ask AutoCAD to keep it as a polyline or as a region, or not to keep it at all:

- In order for the Hatch command to work, it needs to analyze all the objects in the current viewport (in Model space this means the area you are seeing right now), which may take a long time (depending on the number of objects), but you can ask AutoCAD to analyze only the relative objects rather than all objects. Locate the **Select New Boundary Set** button:

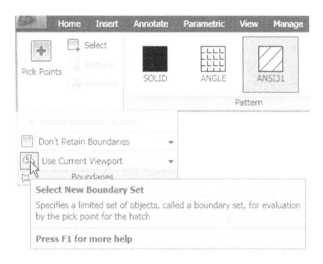

- You will see the following prompt:

```
Select objects:
```

- Select the desired objects, then press [Enter], and this time the pop-up list will read **Use Boundary Set**, instead of the default prompt:

6.14 EDITING HATCHES

- Editing hatches in AutoCAD has never been easier. There are two ways to edit a hatch: by clicking (single-click) or by using Properties (double-click).
- If you single-click a hatch, three things will take place:
  - You will see a grip (small dot) at the center of the area.
  - The context **Hatch Editor** tab will appear, which includes the same panels the Hatch Creation tab does.
  - The Quick Properties palette will appear.
- If you move your mouse to the grip (without clicking) you will see the following menu:

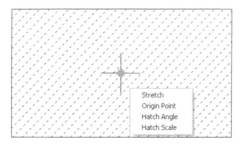

- You can do any or all of the following:
  - Stretch in order to stretch the hatch; however, it is preferable to display boundary objects as discussed above.
  - Modify the Origin Point on the spot.
  - Modify the Hatch Angle on the spot.
  - Modify the Hatch Scale on the spot.
- Also, the context tab Hatch Editor will allow you to make all the necessary modifications because it contains the same panels as the Hatch Creation tab.
- The **Quick Properties** palette will appear as well, and you can make any needed edits:

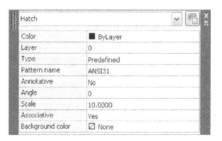

- On the other hand, double-clicking the desired hatch, or selecting the hatch, right-clicking, and selecting the **Properties** option will show the **Properties** palette:

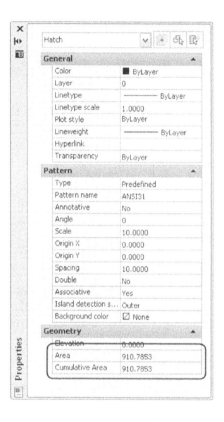

- The data valid for editing is everything related to the hatch properties, options, and boundaries. The Properties palette also provides a single piece of information that other methods don't: the Area of the hatch. If you select a single hatch (whether a single area or multiple areas hatched as a single area) the **Area** field and the **Cumulative Area** will be the same. But if you select multiple areas created using different commands, only the **Cumulative Area** will be filled and the **Area** field showing **Varies** will have a value.
- If you select the hatch and right-click, you will see some Hatch-related editing commands:

- These commands include:
  - **Hatch Edit**, which will show the old Hatch dialog box.
  - **Set Origin**, which will help you set a new origin.
  - **Set Boundary**, which will help you set a new boundary set.
  - **Generate Boundary**, which will regenerate a new boundary for this hatch.

6.15 HATCHES AND TOOL PALETTES

- Just like blocks, you can store hatches in Tool Palettes. Storing hatches in Tool Palettes is not as important as storing hatches with altered properties, such as layer, scale, color, etc., to ensure simple and fast hatch insertion in the drawing. Go to the **View** tab, locate the **Palettes** panel, and select the **Tool Palettes** button:

- You will see Tool Palettes displayed; choose the Hatches tab, just like the following:

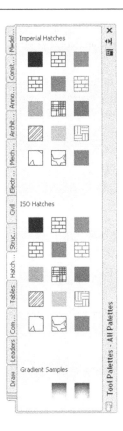

- Drag-and-drop from the drawing and to the drawing. Inside the tool palette right-click and choose the **Properties** option to change the properties of the saved hatch.

HATCH BOUNDARIES AND HATCH EDITING

 Practice 6-7

1. Start AutoCAD 2012.
2. Open **Practice 6-7.dwg**.
3. Select the existing hatch at the top-left of the drawing.
4. From the Boundaries panel, click Recreate, then select Polyline, and Yes. You can see that AutoCAD recreated a new boundary for this orphan hatch.
5. Using the existing hatch at the middle left, select the hatch so you can see the special grip, and then change the angle to 90°.

6. Using the existing hatch at the bottom left, select the hatch so you can see the special grip, and then change the origin point to the lower-left corner and the scale to = 0.1

7. Using the existing hatch at the upper right, select and right-click, then select the **Properties** option; change the scale to 0.5 and the background color to Magenta. What is the total area of the hatch? _____

8. Start the Hatch command, go to the Options panel, and change Island Detection to Normal. Close the Hatch command without doing anything. Start Tool Palettes and go to the Hatches tool palette. Under Imperial Hatches, locate Steel, make another copy of the Steel hatch, and change the scale to = 2.0, then drag it and drop it in the middle-right shape.

9. Using the newly created hatch in the tool palette, drag-and-drop to the last shape left. Select the hatch, and using the Boundaries panel, select Display Boundary Objects. The boundary will display. Select the middle rectangle at the right portion of the polyline and drag to the right for a short distance. What happened to the hatch?

_____

10. Save and close the file.

NOTES:

CHAPTER REVIEW

1. The command to convert a block to a file is:
   a. Makeblock
   b. Createblock
   c. Wblock
   d. None of the above
2. There are _____ hatch types in AutoCAD.
3. Tool Palettes can store commands such as line, polyline, etc.
   a. True
   b. False
4. _____ is the tool that allows you to share blocks, layers, etc.
   a. Tool Palettes
   b. Design Office
   c. Design Center
   d. The Internet
5. Using Tool Palettes you can customize all tools by changing their properties.
   a. True
   b. False
6. Double-clicking a block will show the _____ dialog box.
7. Hatch grips are like regular grips.
   a. True
   b. False
8. If both Design Center and Tool Palettes are open, you can drag any block from Design Center to a tool palette.
   a. True
   b. False

CHAPTER REVIEW ANSWERS

1. c
3. a
5. a
7. b

7 TEXT AND TABLES

7.1 CREATING TEXT AND TABLES

- Inserting text in AutoCAD involves two simple steps:
  - Creating a text style (normally created only once) that includes the size and the shape of the text.
  - Inserting text using either Single Line Text or Multiline Text (it looks and acts like text in Microsoft Word).
- Normally, creating a style is tedious and lengthy and is usually the responsibility of CAD managers as part of company standardization.
- Table styles are part of this standardization and include:
  - Creating a table style.
  - Inserting and filling a table.
- Text and table styles and all other styles should be a part of the document template that holds company standards. If you don't have a template, you can still share styles using Design Center (as discussed in Chapter 6).

7.2 HOW TO CREATE A TEXT STYLE

- The first step in adding text in AutoCAD is to create a text style, which is where you define the characteristics of your text. To start the Text Style command, go to the **Annotate** tab, locate the **Text** panel, and click the small arrow at the lower right:

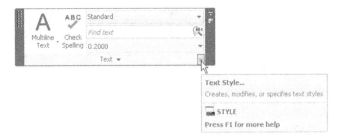

- The following dialog box will appear:

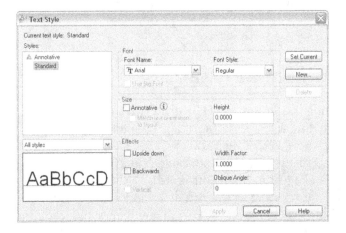

- There are two pre-defined text styles; one is called **Standard** and the other is called **Annotative**. They are almost the same, except the latter uses the Annotative feature (which will be discussed in Chapter 9). Both styles use the Arial font. Professional users usually create their own text styles. To create a new text style, click the **New** button, and you will see the following dialog box:

- Type the name of the new text style and click **OK**. The first thing you will do is select the **Font**. There are two types of fonts you can use in AutoCAD:
  - Shape files (*.shx), the very old method of fonts (out-of-date).
  - True Type Fonts (*.ttf).
- See the illustration below and notice how .ttf files are more accurate and look better:

True Type Font   Shape File Font

- Next, you will select the **Font Style** (if you selected a true type font), and you will choose one of the following:
  - Regular
  - Bold
  - Bold Italic
  - Italic
- Keep **Annotative** off for now (we will discuss it in Chapter 9). Next, you have to specify the **Height** of the text, which is the height of the capital letters (lowercase letters will be two-thirds of the height). See the following illustration:

Baseline
Text height is the capital letter height

- You have two options when setting text height:
  - You can set the height to be 0 (zero), which means the height will be variable (you have to input the value each time you use this style).
  - You can set the height to a value greater than 0 (zero), which means the height is fixed.
- Finally, you will set the effects. You have five of them:
  - Upside down, to write text upside down.
  - Backwards, to write text from right to left.

- Vertical, to write text from top-to-bottom. Good for Chinese words, but only for Shape files.
- Width Factor, to set the relationship between the width of the letter and its length. If the value is >1.0, then the text will be wide. If the value is <1.0, the text will be long.
- Oblique Angle, to set the angle to italicize either to the right (positive value) or to the left (negative value).
- When you are done, click the **Apply** button, and then **Close**.
- You can show **All styles** or show **Styles in use**. See the pop-up list at the left as shown below:

7.3 SINGLE LINE TEXT

- This command will allow you to create single lines of text, and each line will be independent of the other lines. To issue this command, go to the **Annotate** tab, locate the **Text** panel, and then click the **Single Line** button:

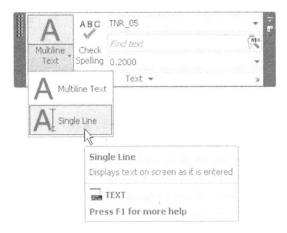

- You will see the following prompts:

```
Current text style: "TNR_05" Text height: 0.5000
Annotative:  No
Specify start point of text or [Justify/Style]:
Specify rotation angle of text <0>:
```

- The first prompt gives you some information about the current settings, the current style (in our example TNR_05) and the current height (in our example = 0.50) and whether this text will be Annotative or not.
- At the first prompt you can change the current Justification (discussed in the section on Multiline Text) and Style settings by typing **J** or **S**, or you can specify the start point of the baseline of the text, then specify the rotation angle (default value is 0 (zero)). Once you press [Enter] you will see a blinking cursor ready for you to type any text you want. To finish any line, press [Enter], and to end the command press [Enter] twice.
- Another way of setting the current text style is to go to the **Annotate** tab, then locate the **Text** panel, and you will you see at the top-right the current text style, which you can change:

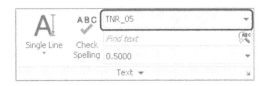

CREATING A TEXT STYLE AND SINGLE LINE TEXT

 Practice 7-1

1. Start AutoCAD 2012.
2. Open **Practice 7-1.dwg**.
3. Create a text style with the following specs:
   a. Name = Room Names
   b. Font = Tahoma
   c. Annotative = off
   d. Height = 0.3
   e. Leave the rest as default values
4. Make the Room Names text style current.
5. Make layer Text current.
6. Type the room names as shown below.
7. Make the Standard text style current (with this text style the height = 0, so you should set it every time you want to use this style).
8. Make layer Centerlines current.
9. Zoom to the upper-left centerline, and notice that the letter A is missing.
10. Start Single Line Text, right-click, and select the Justify option. From the list choose MC (Middle Center), select the center of the circle as the Start point, for the Height set it to 0.25, Rotation angle = 0, then type A, and press [Enter] twice.
11. You should get the following results:

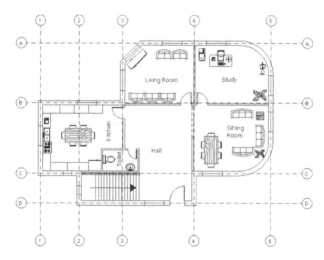

12. Save and close the file.

7.4 MULTILINE TEXT

- This command will allow you to type text in an environment similar to Microsoft Word or other word processor. To issue this command, go to the **Annotate** tab, locate the **Text** panel, and then click the **Multiline Text** button:

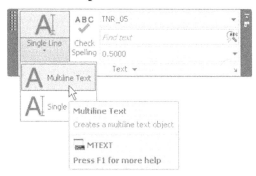

- You will see something like the following in the Command window:

```
Current text style: "TNR_05" Text height: 0.500
Annotative No
Specify first corner:
Specify opposite corner or [Height/Justify/Line
spacing/Rotation/Style/Width]:
```

- The first prompt gives you some information about the current settings, including the current style (in our example TNR_05) and the current height (in our example =0.50) and whether this text will be Annotative or not. You will need to specify an area to write in, so the cursor will change to:

- At the Command window, you will see the following prompt:

```
Specify first corner:
```

- Specify the first corner, and you will see something like the following:

- At the Command window you will see the following prompt:

```
Specify opposite corner or [Height/Justify/Line
spacing/Rotation/Style/Width/Columns]:
```

- Specify the other corner, or select one of the options, and accordingly the Text Editor with a ruler will appear:

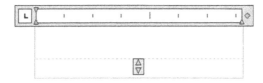

- AutoCAD will show a context tab called **Text Editor**, which looks like the following:

- These panels will allow you to do many things; they are discussed as follows.

7.4.1 Style Panel

- This is the Style panel:

- Use the Style panel to select the text style you want to use and set the height (this value will overwrite the text style height, so be careful).

7.4.2 Formatting Panel

- The following shows the Formatting panel:

- Use the **Formatting** panel to do all or any of the following:
  - Make the selected text **Bold**.
  - Make the selected text **Italic**.
  - Make the selected text **Underlined**.
  - Make the selected text **Overlined**.
  - Change the **Font** of the selected text.
  - Change the **Color** of the selected text.
  - Change capital letters to lowercase letters and vice versa.
  - Change the **Background Mask** for the selected text, and you will see the following dialog box:

  - Change the **Oblique Angle** of the selected text.
  - Change the **Tracking** (to increase or decrease the spaces between letters. (If the value is greater than 1 this means there will be more spaces between letters, and vice versa).
  - Change the **Width Factor.**

7.4.3 Paragraph Panel

- The following shows the Paragraph panel:

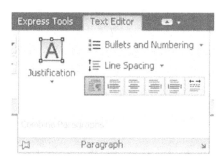

- Use the Paragraph panel to do all or any of the following:
    - Change the **Justification** of the text relative to the text area selected; choose one of the following options:

- The following is an illustration of the nine points available relative to the text area:

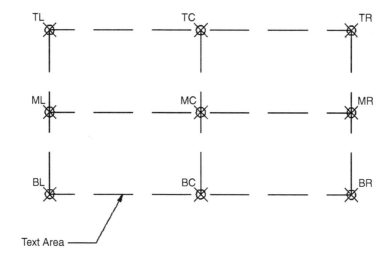

Text Area

- Change the text selected to use **Bullets and Numbering**; there are three choices (Numbered, Lettered, Bulleted):

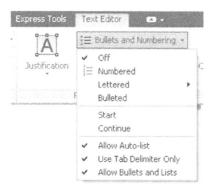

- Change the **Line Spacing** of the paragraph:

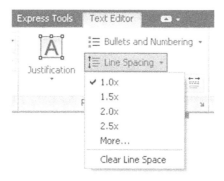

- Change the horizontal justification of the text by using one of the six buttons as shown below:

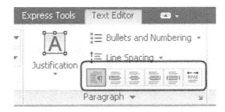

7.4.4 Insert Panel

- The following shows the Insert panel:

- In the Insert Panel, you can do all or any of the following to the selected text:
    - Convert text to two **Columns** or more. If you click the **Columns** button, the following menu will appear:

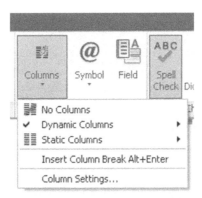

    - Use **Dynamic Columns** to select whether you want AutoCAD to specify the height (Auto height) or you want to set the height (Manual height). See the illustration below:

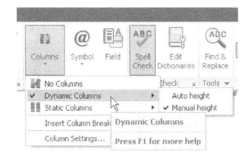

- Select the **Static Columns** option to specify the number of columns:

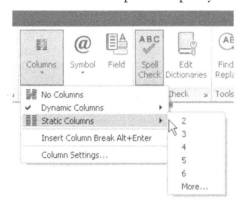

- Select the **Insert Column Break Alt+Enter** option to insert a column break at a certain line, which means the rest of the column will go to the next column.
- Select the **Column Settings** option to show the **Column Settings** dialog box, which will allow you to adjust all of the above settings as well as set the **Column** width and **Gutter** distance:

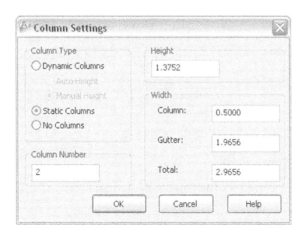

- Here is an example:

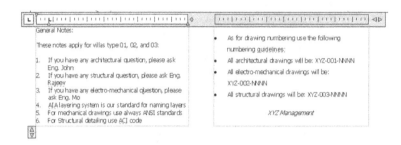

- Select **Symbol** to add scientific characters to your text. You will see the 20 available symbols:

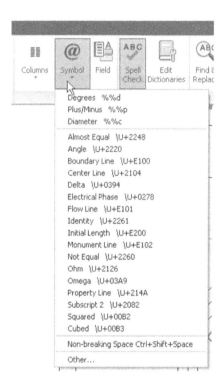

7.4.5 Spell Check Panel

- The following shows the Spell Check panel:

- As long as you are in the Text Editor, you can leave the **Spell Check** button on to catch any misspelled words. You will see a dotted red line underneath the misspelled words, as shown in the following example:

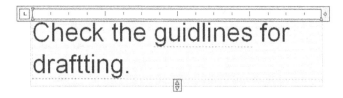

- To get suggestions for the correct word, move to the word, and right-click, and you will see something like the following:

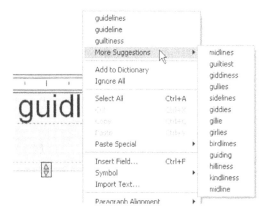

- Select the **Edit Dictionaries** button and you can select another dictionary other than the default one.

7.4.6 Tools Panel

- The following shows the Tools panel:

- In the **Tools** panel, you can do all or any of the following:
  - Click the **Find & Replace** button to search for a word and then replace it with another word. You will see the following dialog box:

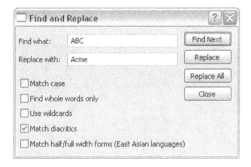

  - Use the **Import Text** button to import text from text files.
  - Use the **AutoCAPS** button to type in the Text Editor using capital letters.

7.4.7 Options Panel

- The following shows the Options panel:

- Using the Options panel you can do all or any of the following:

- Change the **Character Set**, change the **Editor Settings**, or learn more about Multiline Text through **Help**:

- Show the ruler (by default it is displayed) or hide it.
- Undo and redo text actions.

7.4.8 Close Panel

- The following shows the Close panel, which contains a single button to close the Text Editor and all its panels:

7.4.9 While You Are in the Text Editor

- While you are in the Text Editor you can do the following things:
  - Using the ruler you can set the First line indent and the Paragraph indent:

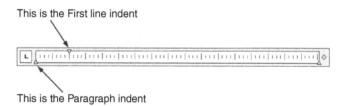

  - If you right-click you will see the following menu, which includes all the functions discussed above:

- You can change the width and height of the area using the following controls:

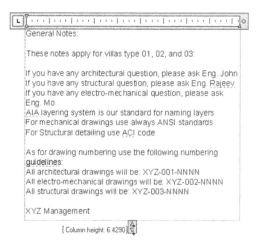

MULTILINE TEXT

   Practice 7-2

1. Start AutoCAD 2012.
2. Open **Practice 7-2.dwg**.
3. You are now in the layout called Cover.
4. Make sure that the current layer is Text.
5. Make the text style Title the current text style.
6. Start Multiline Text, and specify the two corners of the rectangle as your text area.
7. Before you type anything, change the Justification to MC.
8. Type the following as shown:

9. Erase the rectangle.
10. Go to the layout called ISO A1 – Overall.

11. Make the Notes text style the current text style.
12. Using the rectangle at the right, specify the two corners of your text area.
13. Using the Tools panel, select the Import Text command, and select the Notes.txt file.
14. Select General Notes and make it Bold, Underlined, and size = 5.0.
15. Stretch the width a little bit so 03 will be on the same line.
16. Select the text from the first line, "If you have…" to "ACI code," then select Bullets and Numbering and select Numbered.
17. Select the text starting from, "As for…" until the last "NNNN," and make it Bulleted and Line spacing = 1.5 x.
18. Select the last line "XYZ Management" and make it Centered and Italic.
19. Close the Text Editor.
20. Erase the rectangle.
21. Go to ISO A1 – Architectural Details.
22. Using the rectangle, start Multiline Text, select two opposite corners, and import the same file Notes.txt.
23. Do the same thing you did above and make the text Numbered and Bulleted.
24. Start the Column command and set two static columns, then using the Column Settings dialog box, set Column Width =125 and Gutter = 22.5.
25. Using the two arrows in the lower-left corner, make sure that all six numbers are in the first column.
26. Erase the rectangle.
27. This what you should have:

General Notes:

These notes apply for villas type 01, 02, and 03:

1. If you have any architectural question, please ask Eng. John
2. If you have any structural question, please ask Eng. Rajeev
3. If you have any electro-mechanical question, please ask Eng. Mo
4. AIA layering system is our standard for naming layers
5. For mechanical drawings use always ANSI standards
6. For Structural detailing use ACI code

• As for drawing numbering use the following numbering guidelines:
• All architectural drawings will be: XYZ-001-NNNN
• All electro-mechanical drawings will be: XYZ-002-NNNN
• All structural drawings will be: XYZ-003-NNNN

XYZ Management

28. Save and close the file.

7.5 TEXT EDITING

- There are several ways to edit text including double-clicking and using Quick Properties, Properties, and Grips.

7.5.1 Double-clicking Text

- To edit text whether it is Single Line Text or Multiline Text simply double-click it. If it is Single Line Text, you will see the text selected and be able to add to it or modify it. If it is Multiline Text, you will see the Text Editor again and the Text Edit context tab will appear at the top where you can make all your changes.

7.5.2 Quick Properties and Properties

- To show the Quick Properties for either Single Line Text or Multiline Text, simply click the desired text. You will see something like the following (Text in the following illustrations is Single Line Text and MText is Multiline Text):

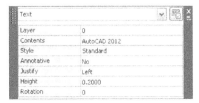 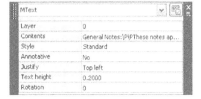

- You can change the following settings: Layer, Contents, Style, Annotative (Yes, or No), Justification, Height, and Rotation angle.
- If you select Single Line Text or Multiline Text and right-click, then select the Properties option, you will see something like the following (again, Text is Single Line Text and MText is Multiline Text):

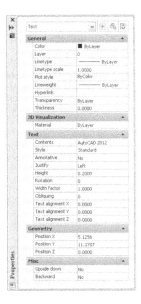 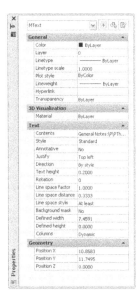

- As you can see, you can change anything related to the selected text.

7.5.3 Editing Using Grips

- You will see a grip at the start point of the baseline and another one at the Justification point (you may see only one point if both points coincide) if you click Single Line Text:

Kitchen

- You will see a single grip appear at the Justification point selected when Multiline Text is selected (in the following example the Justification point is TL), and then you will see two triangles, one at the lower part of the text and one at the right side. The lower triangle will allow you to cut your Multiline Text into columns; simply stretch it up and the text will be cut into two columns. The triangle on the right will allow you to increase/decrease the horizontal distance of the text, hence increasing/decreasing the height of the text:

These notes apply for villas type 01, 02, and 03:

- If you click any multi-column text, you will see the same thing for the first column at the left, except the arrow at the bottom will be pointing downward instead of upward. For the other columns you will see an arrow at the right side. For the last column on the right, you will see an extra arrow at the right, which will allow you to increase/decrease the width and height of the whole text in one shot. See the following illustration:

7.6 SPELL CHECK AND FIND AND REPLACE

- While you are in the Text Editor, you can use Spell Check and Find and Replace, but what if you aren't in the Text Editor? You can still use these features. AutoCAD can spell check the whole drawing and not just the current text; AutoCAD can also spell check the current space and/or layout (layouts will be discussed in Chapter 9) or any selected text. To issue the Check Spelling command, go to the **Annotate** tab, locate the **Text** panel, and click the **Check Spelling** button:

- You will see the following dialog box:

- If AutoCAD finds any misspelled word, it will give you suggestions to choose from or you can simply ignore the misspelled word altogether. AutoCAD can also find any word or part of a word and replace it in the entire drawing file.
- To issue the **Find and Replace** command, go to the **Annotate** tab, locate the **Text** panel, and then type the word you want to replace in the *Find text* field, as shown below, then click the small button at the right:

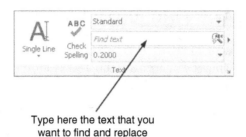

Type here the text that you
want to find and replace

- You will see the following dialog box:

- Under **Replace with**, type the new word(s) you want. Select one of the following choices: **Find**, **Replace**, and **Replace All**. When done, click **Done**.

EDITING TEXT

 Practice 7-3

1. Start AutoCAD 2012.
2. Open **Practice 7-3.dwg**.
3. You are now in the Cover layout. Click the text, using the arrow at the bottom, and drag it upward to cut the text into two columns, each holding two lines.
4. Use the arrow at the right of the second column and drag it until the two columns touch each other. Press [Esc] to end the editing process.
5. Go to layout ISO A1 – Overall.
6. Zoom to the text at the right, select it, right-click, and choose the Properties option. Change the following:
   a. Change the style from Notes to Standard
   b. Change Justify to Middle Center
   c. Change Text Height to 4.5

7. Using the arrow at the right, stretch the text to the right by 10 units (if OSNAP is annoying you switch it off).

8. Press [Esc] to end the editing process.

9. Go to the Annotate tab, locate the Text panel, and in the Find and Replace field, type XYZ, and then click the small button at the right. When the dialog box comes up, in the Replace with field, type ACME, and then click Replace All.

10. Go to layout ISO A1 Architectural Details, double-click the multi-column text, and convert it to single column text.

11. Press [Esc] to end the editing process.

12. Go to Model Space.

13. Click the word "Hall," and the Quick Properties will appear. Set the Justification to Middle Center. Using the Move command try to put this word in the middle center of the space.

14. Save and close the file.

7.7 CREATING A TABLE STYLE

- In order to create a professional table, you should create a table style, which holds the features of the table and specifies the Title, Header, and the Data rows. Using this style you can insert as many tables as you wish. Table styles can be shared using Design Center.
- To issue the command, go to the **Annotate** tab, locate the **Tables** panel, and then select the **Table Style** button:

- The following dialog box will be displayed:

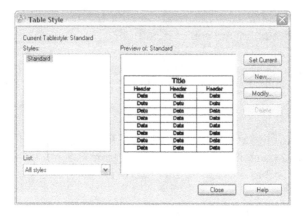

- Just like for text styles, there is a pre-defined table style called Standard. Click the **New** button to create a new table style, and you will see the following dialog box:

- Input the name of the new table style, and click **Continue**.
- You will see the following dialog box:

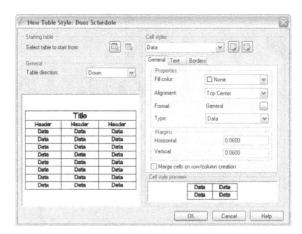

- AutoCAD will allow you to start a new table style based on an existing table in your current drawing; click the button shown here:

- If you want a new style, you need to specify the characteristics of your table by selecting the **Table direction**, **Down** or **Up**; see the following illustration:

- AutoCAD creates three parts for any table: Title, Header, and Data. Using the Table Style command you can set these three parts by selecting the desired part, then using the **General** tab, **Text** tab, and the **Border** tab to create your settings:

7.7.1 General Tab

- The General tab is shown below:

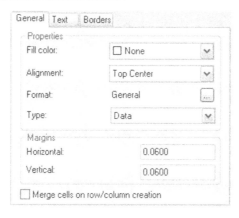

- Edit the following features:
  - Change the **Fill Color** of the cells (by default it is None).
  - Change the **Alignment** to set up the text relative to the cell borders. For instance, if you choose Top Left, the text will reside in the top-left part of the cell. See the following illustration:

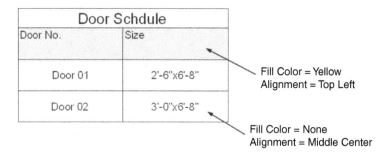

- Change the **Format**. You will see the following dialog box, which will allow you to set the format of the cell, whether it is currency, percentage, date, etc.:

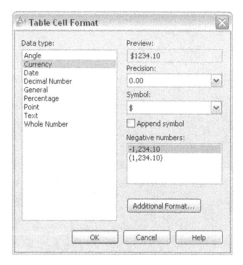

- Change the **Type** of the contents of the cell to **Data** or **Label**. This part is very important, since some of the cells may hold numbers, but these numbers shouldn't be included in a mathematical formula, so the type of data will be Label and not Data.
- Under **Margins**, control the **Horizontal** and **Vertical** distances around the Data relative to the borders.

7.7.2 Text Tab

- The Text tab is shown below:

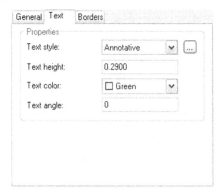

- Edit the following features:
  - Change the **Text style** of the text filling the cells.
  - Change the **Text height** of the text filling the cells (keeping in mind that if the text style has a height > 0.0, the number here is meaningless).
  - Change the **Text color**.
  - Change the **Text angle**; see the following illustration:

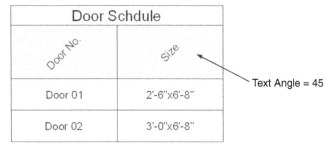

7.7.3 Borders Tab

- The Borders tab is shown below:

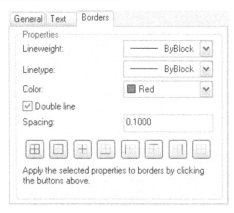

- Edit the following features:
  - Change the **Lineweight**, **Linetype**, and **Color** of the border lines.
  - Change the border to **Double** line instead of **Single** line (default value), and also choose the **Spacing** between the lines.
  - Change whether you want lines representing column separators and row separators.

7.8 INSERTING A TABLE IN THE CURRENT DRAWING

- This command will allow you to insert a table in the current drawing. To issue this command, go to the **Annotate** tab, locate the **Tables** panel, and then select the **Table** button:

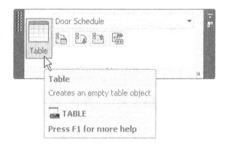

- You will see the following dialog box:

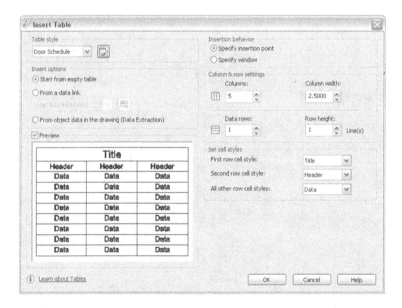

- The first step is to select the desired pre-defined **Table style** name from the available list. If you didn't create a table style yet, click the small button beside the list to define it now.
- Select the proper **Insert options**. To insert the table in your drawing, choose one of the following three options:
  - **Start from an empty table**.
  - **Start from a data link** – this is an advanced feature of AutoCAD.
  - **Start from object data in the drawing (Data Extraction)** – this is also an advanced feature of AutoCAD.
- Most often you will use the first option, **Start from an empty table**. Now select the **Insertion behavior**; there are two available choices:
  - Specify the insertion point.
  - Specify the window.

7.8.1 Specify Insertion Point Option

- The insertion point here is the upper-left corner of the table. You will hold the table from this point. Use the following data for the rest of the table:
  - Columns (the number of columns).
  - Column width.
  - Data rows (the number of rows, but without Title and Heads).
  - Row height (in lines).

- Click **OK**, and AutoCAD will show the following prompt:

```
Specify insertion point:
```

- Specify the location of the table and start filling the cells. You can use the arrows on the keyboard or the [Tab] key to jump from one cell to another. Use [Shift] + [Tab] to go backwards.

7.8.2 Specify Window Option

- Using this option you will specify a window, which means you will give Auto-CAD the total length and the total width of the table. In order to fulfill the rest of the information, input the following:

- The number of **Columns**, and AutoCAD will calculate the column width. Or input the **Column width**, and AutoCAD will calculate the number of columns.
- The number of **Data rows**, and AutoCAD will calculate the row height. Or input the **Row height**, and AutoCAD will calculate the number of rows:

- Click **OK**, and AutoCAD will show the following prompts:

```
Specify first corner:
Specify second corner:
```

- Specify the two opposite corners of the table, then start inputting the cell contents using the arrows, [Tab], and [Shift] + [Tab].

CREATING A TABLE STYLE AND INSERTING A TABLE IN THE CURRENT DRAWING

 Practice 7-4

1. Start AutoCAD 2012.
2. Open **Practice 7-4.dwg**.
3. Create a new table style with the following settings:
   a. Name = Door Schedule
   b. Title Text Style = Notes
   c. Header Fill Color = Yellow
   d. Header Text Style = Notes
   e. Data Alignment = Middle Left
   f. Data Horizontal Margin = 10

g. Data Text Style = Notes

h. Data Text Color = Blue

4 Create a table using the above table style and the frame drawn using the Window option to insert the table:

Door Schedule	
Door No.	Size
Door 01	2'-6" x 6'-8"
Door 02	3'-0" x 6'-8"

5. After finishing the table input, erase the frame.

6. Save and close the file.

NOTES:

CHAPTER REVIEW

1. There are two types of text in AutoCAD, Single Line and Multiline.
   a. True
   b. False
2. In _____ you will define the background color of the table cell.
3. Which one of the following is <u>not</u> a panel in the Text Editor context tab:
   a. Insert
   b. Options
   c. Text
   d. Paragraph
4. There are two methods to insert a table in a drawing.
   a. True
   b. False
5. By default the first row cell style is _____.
6. Height in a text style is for:
   a. Small letters
   b. Capital letters and lowercase letters
   c. Capital letters only
   d. Everything above the baseline
7. When inserting a table using a window, specifying the number of columns is enough; you don't need to specify the column width.
   a. True
   b. False
8. In a table style you can specify different text styles for the Header and Title.
   a. True
   b. False

CHAPTER REVIEW ANSWERS

1. a
3. c
5. Title
7. a

Chapter **8**   **DIMENSIONS**

In This Chapter

◇ Dimensioning and dimension types
◇ How to create a dimension style and sub-style
◇ How to insert different types of dimensions
◇ How to edit dimension blocks
◇ How to create a multileader style
◇ How to insert a multileader dimension

8.1 WHAT IS DIMENSIONING IN AUTOCAD?

- Dimensioning in AutoCAD is just like using text and tables. You should prepare your dimension style first, and then use it to insert dimensions. Dimension styles control the overall outcome of the dimension block generated by the different types of dimension commands.
- To insert a dimension, depending on the type of the dimension, you should specify points, or select objects, and then a dimension block will be added to the drawing. For example, in order to add a linear dimension you will select two points representing the distance to be measured, and a third point will be the location of the dimension block. See the illustration below:

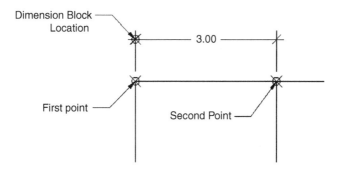

- The generated block consists of three portions; they are:
  - Dimension line.
  - Extension lines.
  - Dimension text.
- See the following illustration:

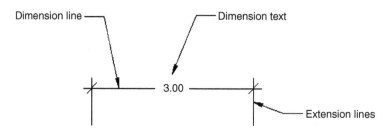

8.2 DIMENSION TYPES

- The following are the dimensions types in AutoCAD:

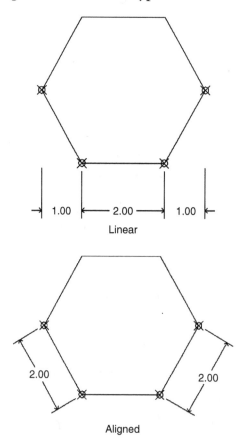

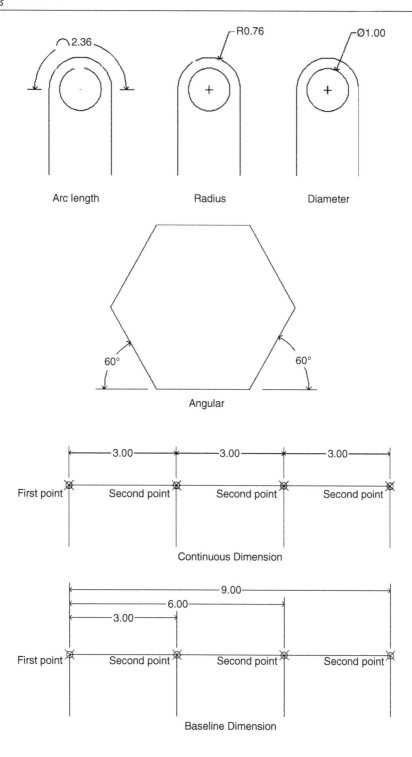

Arc length   Radius   Diameter

Angular

Continuous Dimension

Baseline Dimension

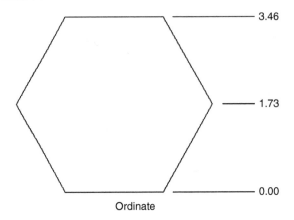

3.46

1.73

0.00

Ordinate

8.3 HOW TO CREATE A NEW DIMENSION STYLE

- This command will allow you to create a new dimension style, or modify an existing one. To issue this command, go to the **Annotate** tab, locate the **Dimensions** panel, and then select the **Dimension Style** button:

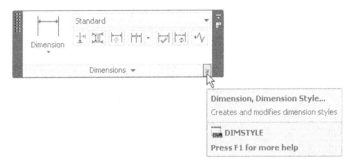

- You will see the following dialog box:

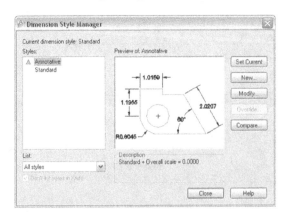

- As you can see there are two predefined dimension styles, Standard and Annotative. Click the **New** button to create a new style. You will see the following dialog box:

- Input the name of the new style. Under **Start With**, select the existing style that you will use as your starting point. Leave the Annotative checkbox off (we will discuss it in Chapter 9). Make sure that under Use for All dimensions is selected (we will cover it at the end of our discussion), and then click the **Continue** button.

8.4 DIMENSION STYLE: LINES TAB

- As a rule-of-thumb, and while we are discussing the different dimension style tabs, we will leave Color, Linetype, and Lineweight at their default settings because we want these things to be controlled by the layer rather than the individual dimension block.
- The first tab in the dimension style dialog box is Lines, and it allows you to control the dimension lines and extension lines:

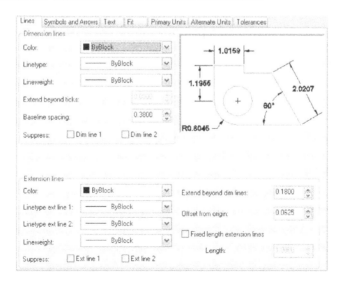

- Under **Dimension lines** change all or any of the following settings:
  - Control **Extended beyond ticks,** which is as illustrated below (this option works only when **Arrowhead** is **Architectural tick** or **Oblique**):

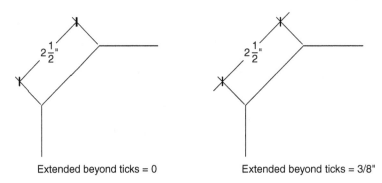

Extended beyond ticks = 0   Extended beyond ticks = 3/8"

- As we will see in this chapter when you add a baseline dimension you will not control the spacing between a dimension and another, you will control the **Baseline spacing**, as illustrated below:

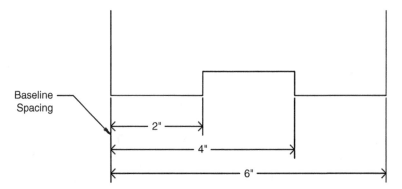

- From now on when we say First, we mean the nearest point to the first point chosen, and Second is the nearest point to the second point chosen.
  - Choose whether to **Suppress Dim line 1, Dim line 2**, or leave them as is. See the illustration below:

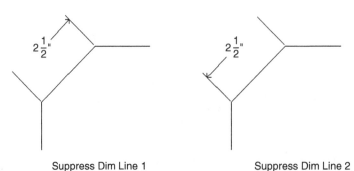

Suppress Dim Line 1   Suppress Dim Line 2

- Under **Extension lines** change all or any of the following settings:
  - Choose whether to **Suppress Ext line 1, Ext line 2**, or leave them as is. See the illustration below:

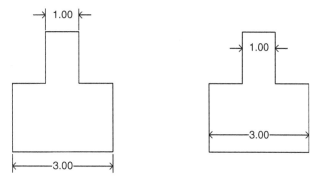

Both extension lines are displayed   Suppress both extension lines

  - Input **Extend beyond dim lines** and **Offset from origin**. See the illustration below:

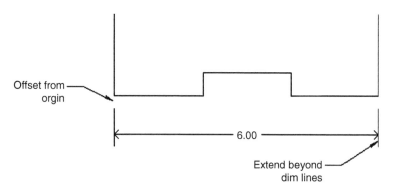

  - Choose whether to fix the length of the extension lines or not, and if yes, the length. See the following example:

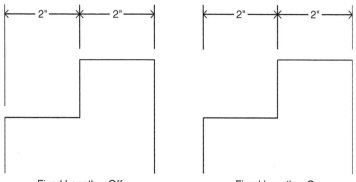

Fixed Length = Off   Fixed Length = On

8.5 DIMENSION STYLE: SYMBOLS AND ARROWS TAB

- This tab will allow you to control the arrowheads and related features. The following shows the **Symbols and Arrows** tab:

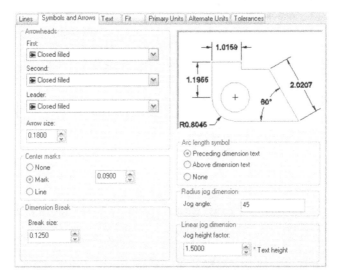

- Under **Arrowheads**, change all or any of the following:
  - The shape of the **First** arrowhead. When you set the shape of the first arrowhead, the **Second** arrowhead will automatically change, but you can also change them manually.
  - The shape of the arrowhead to be used in the **Leader** (Radius and Diameter are not leaders).
  - The **Size** of the arrowhead.
- Under **Center marks**, choose whether to show or hide the center mark of arcs and circles as shown below, then set the **Size** of the center mark:

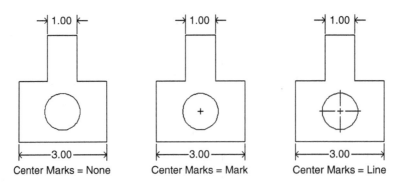

Center Marks = None   Center Marks = Mark   Center Marks = Line

■ Under **Dimension Break**, input the **Break size**. The break size is defined as the distance of the void left between two broken lines of a dimension. See the illustration below:

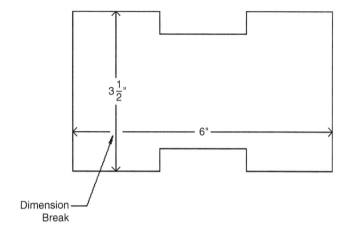

Dimension
Break

■ Under **Arc length symbol**, choose whether to show (as shown in the illustration below) or hide the arc length symbol:

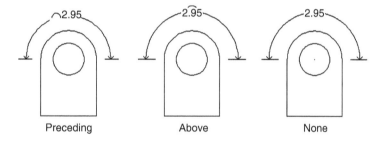

Preceding   Above   None

■ Under **Radius dimension jog**, input the **Jog angle** as shown in the following illustration:

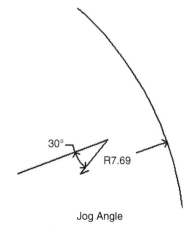

Jog Angle

- Under **Linear jog dimension**, input the **Jog height factor** as shown below. The jog height factor is defined as the factor used to multiply the height of the text used in a dimension:

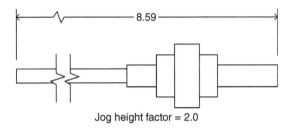

Jog height factor = 2.0

8.6 DIMENSION STYLE: TEXT TAB

- This tab will allow you to control the text appearing in the dimension block. The following is an illustration of the **Text** tab:

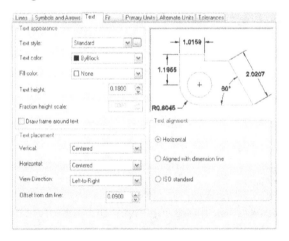

- Under **Text appearance** change all or any of the following:
  - Select the desired **Text style**, or create a new one.
  - Specify the **Text color** and the **Fill color** (text background color):

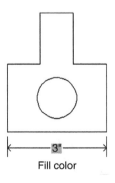

Fill color

- If your text style has a text height = 0 (zero), then input the **Text height**.
- Depending on the primary units (discussed in a moment) set the **Fraction height scale**:

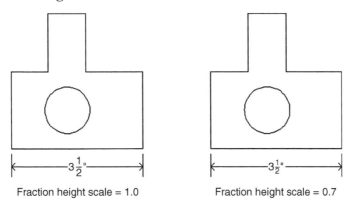

Fraction height scale = 1.0   Fraction height scale = 0.7

- Select whether your text will have a frame. See the illustration below:

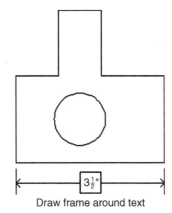

Draw frame around text

- Under **Text placement** change all or any of the following:
  - Choose the **Vertical** placement of your text related to the dimension line. There are five choices: Centered, Above, Outside, JIS (Japan Industrial Standard), and Below. See the following:

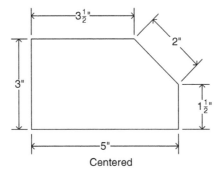

Centered

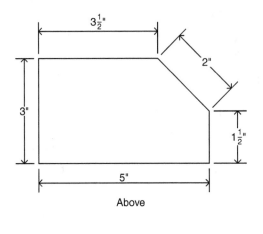

Above

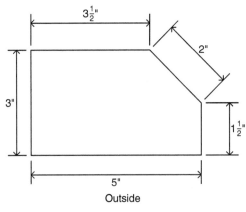

Outside

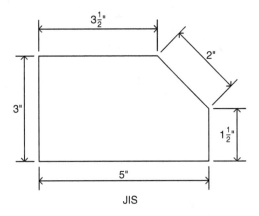

JIS

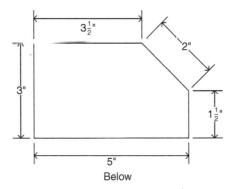

Below

- Choose the **Horizontal** placement. You have five choices to choose from. See the illustration below:

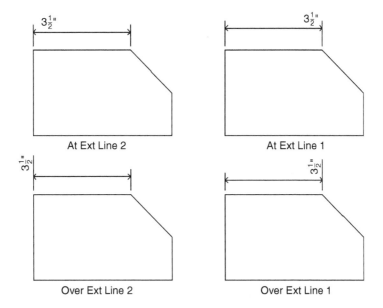

At Ext Line 2   At Ext Line 1

Over Ext Line 2   Over Ext Line 1

- Choose the **View Direction** of the dimension text: Left-to-Right or Right-to-Left. See the following illustration:

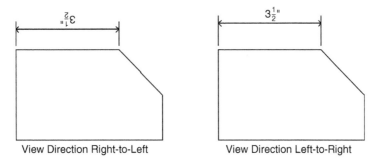

View Direction Right-to-Left   View Direction Left-to-Right

- Input the **Offset from dim line**, as shown below:

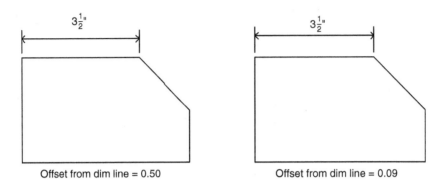

Offset from dim line = 0.50   Offset from dim line = 0.09

- Under **Text alignment**, control the alignment of the text related to the dimension line, whether always horizontal regardless of the alignment of the dimension line, or aligned with the dimension line. ISO will influence only the Radius and Diameter dimensions (all of the dimension types will be aligned except for those two):

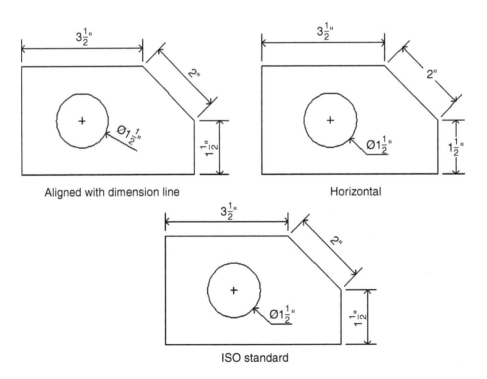

Aligned with dimension line   Horizontal

ISO standard

8.7 DIMENSION STYLE: FIT TAB

- This tab will allow you to control the relationship between the dimension block components. The following shows the **Fit** tab:

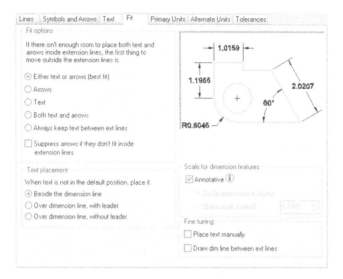

- Under **Fit options**, change all or any of the following:
  - If there is no room for the text and/or the arrowheads inside the extension lines, what do you want AutoCAD to do? Select the proper option.
  - If there is no room for the arrows to be inside the extension lines, do you want AutoCAD to suppress them?
- Under **Text placement**, when the text is not in the default position, select one these options:

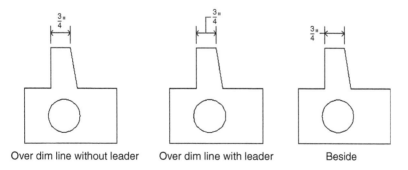

Over dim line without leader   Over dim line with leader   Beside

- Under **Scale for dimension features**, you will control the size of the text (length, size, etc.). Will it be scaled automatically if it is Annotative? Or will it follow the viewport scale if it was input in the layout? If you want to input it in Model space, you can set the scaling factor.
- Under **Fine tuning**, select whether you want to place your text manually or leave it to AutoCAD. Also select whether to always force text inside extension lines.

8.8 DIMENSION STYLE: PRIMARY UNITS TAB

- In this tab, you control everything related to the numbers that appear at the dimension block. The following shows the **Primary Units** tab:

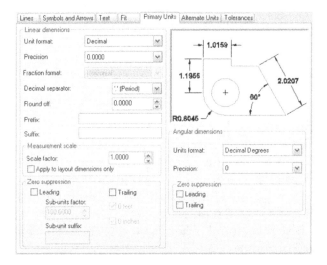

- Under **Linear dimensions**, change all or any of the following settings:
  - Select the desired **Unit format**, then select its **Precision**.
  - If your selection was **Architectural** or **Fractional**, choose the desired **Fraction format**. You have three choices to choose from: **Horizontal**, **Diagonal**, and **Not Stacked**:

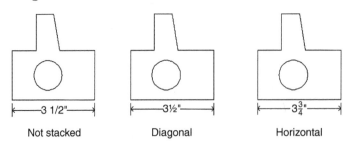

     Not stacked   Diagonal   Horizontal

- If your selection is **Decimal**, then choose the **Decimal Separator**. You have three choices: **Period**, **Comma**, and **Space.**
- Input the **Round off** number.
- Input the **Prefix** and/or the **Suffix**, as shown below:

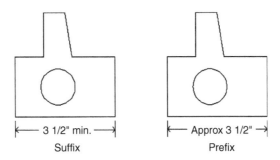

Suffix   Prefix

- Under **Measurement scale**, change all or any of the following:
  - By default AutoCAD will measure the distance between the two points specified by the user (if it is linear) and input the text in the format set. But what if you want to show a different value than the measured value? Then you would input the **Scale factor**.
  - Choose whether this scale will affect only the dimension input in the layout. (We will discuss layouts in Chapter 9.)
- Under **Zero suppression**, choose whether to suppress the **Leading** and/or the **Trailing** zeros as shown below:

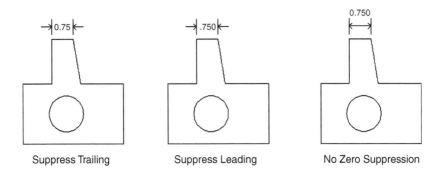

Suppress Trailing   Suppress Leading   No Zero Suppression

- If you have meters as your unit, and the measured value is less than one, this is a sub-unit. Select the sub-unit factor and the suffix for it (in this example it is cm.)
- Under **Angular dimensions**, choose the **Units format** and the **Precision**. Control **Zero suppression** for angles as well.

8.9 DIMENSION STYLE: ALTERNATE UNITS TAB

- This tab will allow you to show two numbers in the same dimension block; one showing the primary units and the other showing the alternate units. The following shows the **Alternate Units** tab:

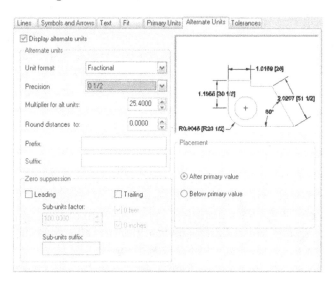

- Click on the **Display alternate units** option, then change the following:
  - Choose the Alternate **Unit format** and its **Precision**.
  - Input the **Multiplier** for all unit values.
  - Input the **Round** distance.
  - Input the **Prefix** and the **Suffix**.
  - Input the **Zero suppression** method.
  - Choose the method of displaying alternate units: **After primary value** or **Below primary value**. See below:

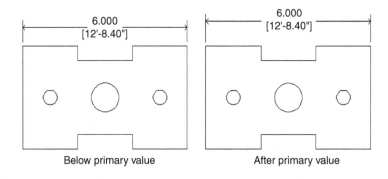

Below primary value   After primary value

8.10 DIMENSION STYLE: TOLERANCES TAB

- This tab will allow you to control whether or not to show tolerances and what method to use. The following shows the **Tolerances** tab:

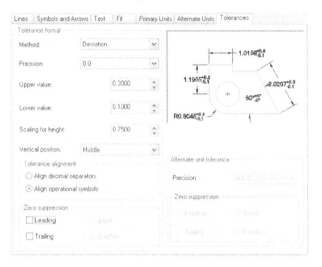

- There are four tolerance formats:
  - Symmetrical.
  - Deviation.
  - Limits.
  - Basic.
- The following is an illustration for each of the four choices:

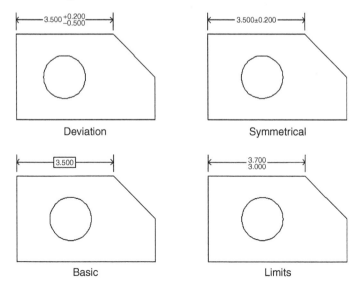

- Under **Tolerance format**, change all or any of the following:
  - Select the proper **Method**, and then select its **Precision**.
  - Depending on the method specify the **Upper value** and **Lower value**.
  - Input **Scaling for height** for the tolerance values if desired.
  - Choose the **Bottom**, **Middle**, or **Top** vertical position for the dimension text with reference to the tolerance values. See the following illustration:

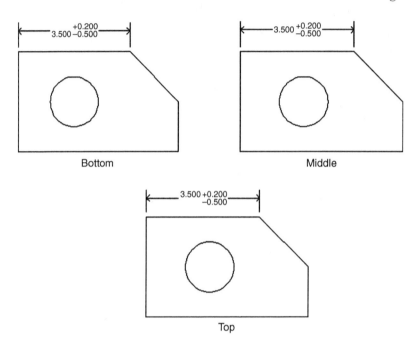

- If the **Deviation** method or **Limits** method is selected, then you have to choose whether to **Align decimal separators** or **Align operational symbols**, as shown below:

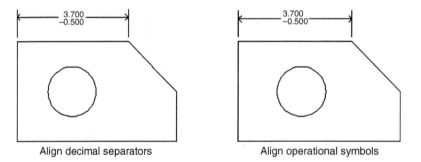

- Under **Alternate unit tolerance**, if the **Alternate units** option is turned on, then specify the **Precision** of the numbers. Consequently, choose **Zero suppression** for both the **Primary units** tolerance and the **Alternate units** tolerance.

8.11 CREATING A DIMENSION SUB-STYLE

- By default the dimension style you create will affect all types of dimensions. If you want the dimension style to affect only a certain type of dimension and not the others you have to create a sub-style. Take the following steps:
  - Select an existing dimension style.
  - Use the **New** button to create a new style, and you will see the following dialog box:

- Go to **Use for** and select the type of dimension (in the following example we selected Diameter) and the dialog box will change to:

- Click the **Continue** button, and make the changes you want; these changes will affect diameter dimensions only.

- The **Dimension Style** dialog box will allow you to differentiate between the style and the sub-style and its features:

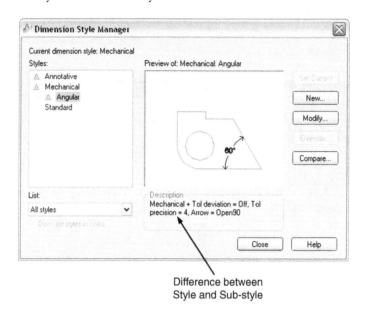

Difference between
Style and Sub-style

CREATING DIMENSION STYLES

 Practice 8-1

1. Start AutoCAD 2012.
2. Open **Practice 8-1.dwg**.
3. Create a new dimension style based on Standard, using the following information:
   a. Name: Part
   b. Extend beyond dim line = 0.3
   c. Offset from origin = 0.15
   d. Arrowhead = Right angle
   e. Arrow size = 0.25
   f. Center mark = Line
   g. Arc length symbol = Above dimension text
   h. Jog angle = 30
   i. Text placement vertical = Above
   j. Offset from dim line = 0.2

      k. Text alignment = ISO Standard
      l. Text placement = Over dimension line with leader
      m. Primary unit format = Fractional
      n. Primary unit precision = 0 ¼
      o. Fraction format = Diagonal
4. Click OK to end the creation process.
5. Select Part, and create a sub-style for Radius, using the following:
      a. Arrowhead = Closed filled
      b. Arrow size = 0.15
6. Save and close the file.

8.12 HOW TO INSERT A LINEAR DIMENSION

- This command will allow you to create a horizontal or vertical dimension. To start the Linear command, go to the **Annotate** tab, locate the **Dimensions** panel, and then select the **Linear** button:

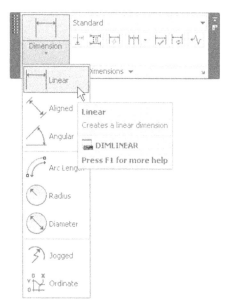

- You will see the following prompts:

```
Specify first extension line origin or <select
object>:
Specify second extension line origin:
Specify dimension line location or
[Mtext/Text/Angle/Horizontal/Vertical/Rotated]:
```

- Specify the first point and second point of the dimension distance to be measured, and then specify the location of the dimension block by specifying the location of the dimension line. The following is the result:

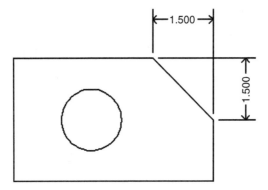

- The prompts also contain other options; they are:
  - Mtext.
  - Text.
  - Angle.
  - Horizontal.
  - Vertical.
  - Rotated.
- Mtext will allow you to edit the measured distance in **MTEXT** mode, while Text will allow you to edit the measured distance in **DTEXT** (Single line) mode. Angle is used to change the angle of the text (the default is 0 (zero)). Horizontal and Vertical will force the dimension to be either horizontal or vertical (the default will allow you to specify either by the movement of the mouse). Finally, Rotated is used to create a dimension line parallel to another angle given by the user.

8.13 HOW TO INSERT AN ALIGNED DIMENSION

- This command will allow you to create a dimension parallel to the two points specified. To start this command, go to the **Annotate** tab, locate the **Dimensions** panel, and then select the **Aligned** button:

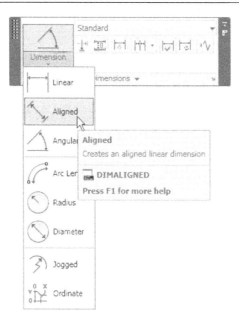

- The following prompts will appear:

```
Specify first extension line origin or <select
object>:
Specify second extension line origin:
Specify dimension line location or
[Mtext/Text/Angle]:
```

- Specify the first point and second point of the dimension distance to be measured, and then specify the location of the dimension block by specifying the location of the dimension line. See the following illustration:

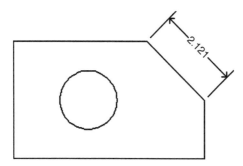

- The rest of the options are identical to the Linear command prompts.

8.14 HOW TO INSERT AN ANGULAR DIMENSION

- This command will allow you to insert an angular dimension between two lines, the included angle of a circular arc, the two points and center of a circle, or three points. To start this command, go to the **Annotate** tab, locate the **Dimensions** panel, and then select the **Angular** button:

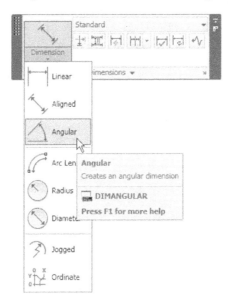

- AutoCAD will use one of the following methods based on the selected objects:
  - If you select a circular arc, it will measure the included angle.
  - If you select a circle, your selected points will be the first point, the center of the circle, and then you will select the third point.
  - If you select a line, it will ask you to select a second line.
  - If you select a point, it will be considered the center point, and it will ask you to specify two more points.

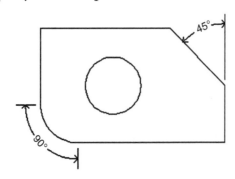

- Based on the above discussion, when you start the command you will see the following prompts (in the following example we selected an arc):

```
Select arc, circle, line, or <specify vertex>:
Specify dimension arc line location or
[Mtext/Text/Angle]:
```

INSERTING LINEAR, ALIGNED, AND ANGULAR DIMENSIONS

 Practice 8-2

1. Start AutoCAD 2012.
2. Open **Practice 8-2.dwg**.
3. Make layer Dimension the current layer.
4. Insert the dimensions as shown below:

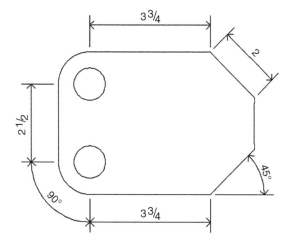

5. Save and close the file.

8.15 HOW TO INSERT AN ARC LENGTH DIMENSION

- This command will allow you to create a dimension measuring the length of an arc. To start this command, go to the **Annotate** tab, locate the **Dimensions** panel, and then select the **Arc Length** button:

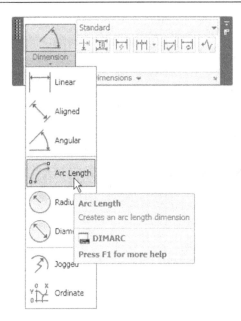

- You will see the following prompts:

```
Select arc or polyline arc segment:
Specify arc length dimension location, or
[Mtext/Text/Angle/Partial/Leader]:
```

- Select the desired arc, and then locate the dimension block, either inside or outside the arc. You will get something like the following:

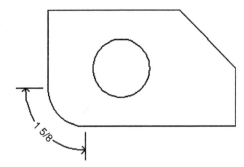

- The options, Mtext, Text, and Angle, were already discussed previously. The Partial option means you want to insert an arc length dimension on part of the arc. You select the arc, then select two internal points on the arc, and you will get something like the following:

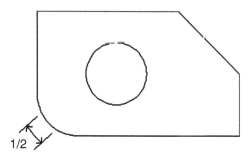

8.16 HOW TO INSERT A RADIUS DIMENSION

- This command will allow you to insert a Radius dimension on an arc or circle. To start this command, go to the **Annotate** tab, locate the **Dimensions** panel, and then select the **Radius** button:

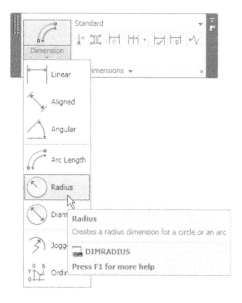

- You will see the following prompts:

```
Select arc or circle:
Specify dimension line location or [Mtext/Text/
Angle]:
```

- Select the desired arc or circle, and then locate the dimension block. You will get something like the following:

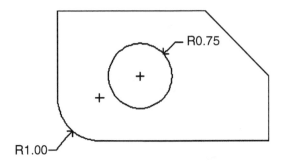

8.17 HOW TO INSERT A DIAMETER DIMENSION

- This command will allow you to insert a diameter dimension on an arc or circle. To start this command, go to the **Annotate** tab, locate the **Dimensions** panel, and then select the **Diameter** button:

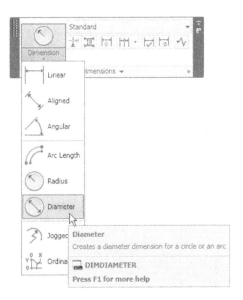

- You will see the following prompts:

```
Select arc or circle:
Specify dimension line location or [Mtext/Text/
Angle]:
```

- Select the desired arc or circle, and then locate the dimension block. You will get something like the following:

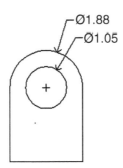

INSERTING ARC LENGTH, RADIUS, AND DIAMETER DIMENSIONS

 Practice 8-3

1. Start AutoCAD 2012.
2. Open **Practice 8-3.dwg**.
3. Make layer Dimension the current layer.
4. Insert the dimensions as shown below:

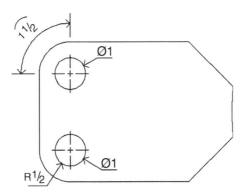

5. Save and close the file.

8.18 HOW TO INSERT A JOGGED DIMENSION

- This command will allow you to insert a jogged arc dimension for a big arc, simulating a new center point. To start this command, go to the **Annotate** tab, locate the **Dimensions** panel, and select the **Jogged** button:

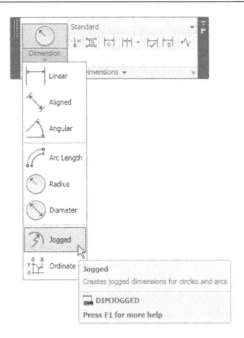

- The following prompts will appear:

```
Select arc or circle:
Specify center location override:
Dimension text = 1.5
Specify dimension line location or [Mtext/Text/
Angle]:
Specify jog location:
```

- As a first step, select an arc or circle, specify the point that will be the new center point (AutoCAD calls it location override), locate the dimension line, and finally, specify the jog's location. You will get something like the following:

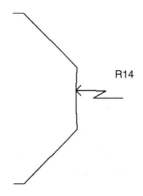

8.19 HOW TO INSERT ORDINATE DIMENSIONS

- This command will allow you to insert dimensions relative to a datum, either in X or in Y. See the following illustration:

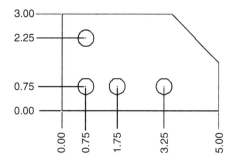

NOTE
- Use the UCS command and the Origin option to relocate the origin to one side of the shape, so the values in both X, Y will be correct. If you leave the origin as the current UCS origin, the values inserted may be wrong.
- To start this command, go to the **Annotate** tab, locate the **Dimensions** panel, and then select the **Ordinate** button:

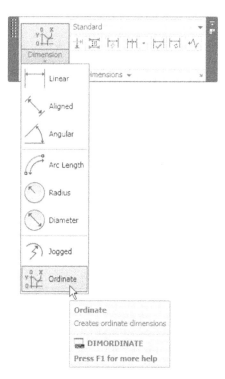

- You will see the following prompts:

```
Specify feature location:
Specify leader endpoint or [Xdatum/Ydatum/Mtext/Text/
Angle]:
```

- First, select the desired point. By default, AutoCAD will give you the freedom to go in the direction of X or in Y. If you want to force the mouse to measure points related to the X axis, then select the **Xdatum** option, and the same applies for the Y axis. We have already discussed the rest of the options.

INSERTING DIMENSIONS

Practice 8-4

1. Start AutoCAD 2012.
2. Open **Practice 8-4.dwg**.
3. Make layer Dimension current.
4. Insert the dimensions as shown below:

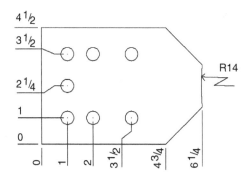

5. Save and close the file.

8.20 INSERTING A SERIES OF DIMENSIONS USING THE CONTINUE COMMAND

- AutoCAD allows you to input a series of dimensions using the Continue command. The Continue command will follow the last dimension command by asking you to input the second point, assuming that the last point of the last dimension will be considered the first point. To start this command, go to the **Annotate** tab, locate the **Dimensions** panel, and then select the **Continue** button.

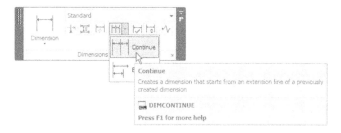

- There are two scenarios when using the Continue command:
  - If there wasn't a dimension command issued in the current AutoCAD session, AutoCAD will ask you to select an existing dimension (linear, ordinate, or angular). You will see the following prompt:

```
Select continued dimension:
```

  - If there was a dimension command issued in the current AutoCAD session, AutoCAD will ask you to continue this command by asking you to specify the second point. You can also select an existing dimension block to continue it, or you can undo the last continue command. You will see the following prompt:

```
Specify a second extension line origin or [Undo/
Select] <Select>:
```

- You will get something like the following:

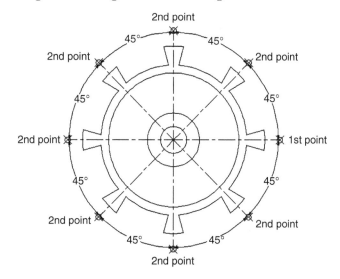

8.21 INSERTING A SERIES OF DIMENSIONS USING THE BASELINE COMMAND

- This command is identical to the Continue command, except all dimensions will be measured based on the first point specified by the user as the baseline. To start this command, go to the **Annotate** tab, locate the **Dimensions** panel, and then select the **Baseline** button:

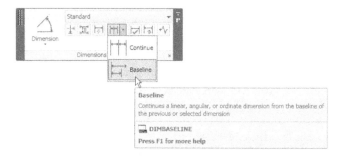

- We don't need to discuss the prompts of this command because they resemble the Continue command prompts. You will get something like the following:

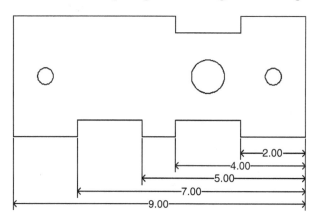

CONTINUE COMMAND

Practice 8-5

1. Start AutoCAD 2012.
2. Open **Practice 8-5.dwg**.

3. Create an angular dimension, and then using the Continue command, complete the shape as follows:

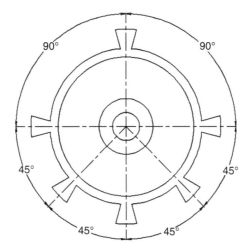

4. Save and close the file.

BASELINE COMMAND

 Practice 8-6

1. Start AutoCAD 2012.
2. Open **Practice 8-6.dwg**.
3. Create a linear dimension, and then using the Baseline command, complete the shape as follows:

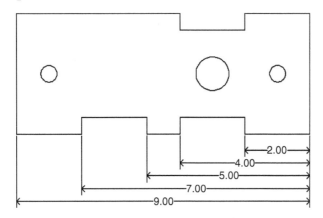

4. Save and close the file.

8.22 USING THE QUICK DIMENSION COMMAND

- This command will allow you to insert a group of dimensions in one shot. To start this command, go to the **Annotate** tab, locate the **Dimensions** panel, and then select the **Quick Dimension** button:

- You will see the following prompts:

```
Associative dimension priority = Endpoint
Select geometry to dimension:
Specify dimension line position, or [Continuous/
Staggered/Baseline/Ordinate/Radius/Diameter/
datumPoint/Edit/settings] <Continuous>:
```

- As a first step, select the desired geometry you want to dimension by using clicking, Window mode, Crossing mode, or any other mode you know. If this is the first time you are using the command in the current AutoCAD session, then AutoCAD will use Continuous as the default option. But if you right-click, you will see the following menu:

- Using this shortcut menu, select the desired dimension type and then specify the dimension line location; consequently, a set of dimensions will be inserted.
- See the following examples:

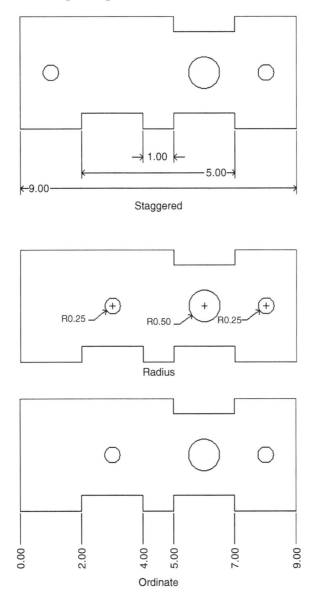

Staggered

Radius

Ordinate

- If you select the Settings option, you will see the following menu:

- This option will allow you to set the default OSNAP for specifying extension line origins.

8.23 EDITING DIMENSION BLOCKS USING GRIPS

- After inserting a dimension block it is easy to edit it using grips, or by right-clicking. Depending on the type of dimension, if you click any dimension block you will see grips in certain places. The following are some examples:
  - For linear and aligned dimensions, grips will appear in five places; at the two points measured, the two ends of the dimension line, and finally, at the dimension text. If you *hover* over the text grip you will see the following:

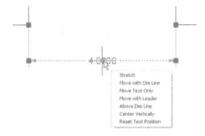

  - AutoCAD will allow you to Stretch, Move with Dim Line, Move Text Only, etc., while holding this grip. If you hover over the two grips at the ends, you will see:

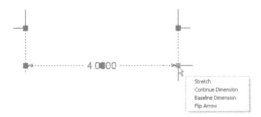

  - Here, you can Stretch and create a continuous or baseline dimension based on the type of dimension selected. You can also flip the arrow nearest to the selected grip.

- Angular dimension grips will appear at five places; at the end points of the two lines involved, at the dimension line, and finally, at the dimension text. If you hover over the text grip you will see the following:

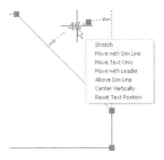

- The commands are identical to linear and aligned dimensions. If you hover over the two end grips you will see the following:

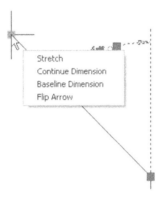

- These are the same commands discussed for linear and aligned dimensions.
- Ordinate dimension grips will appear at four places; the origin point and the measured point, then at the dimension line, and finally, at the dimension text. If you hover over the text grip you will see the following:

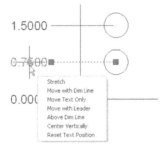

- These are the same commands discussed above. Hovering over the end of the line grip will show the following:

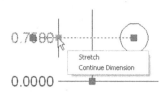

- Ordinate will not work with Baseline, and doesn't have arrows, so these two options are not mentioned for this type of dimension.
- Radius and diameter dimension grips will appear at three places; at the selected point, at the center, and finally, at the dimension text. If you hover over the text grip you will see the following:

- This is the same list of commands. If you hover over the grip at the end of the arrow you will see the following:

- Since radius and diameter don't have the ability to use Continue or Baseline, these two commands are not available. You can use only Stretch and Flip Arrow.
- Arc length will show four grips; one at the two ends of the arc, one at the text, and finally, one near the text. If you hover over the text grip you will see the following:

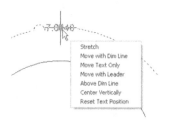

- This is also the same list of commands. If you hover over the end point grip you will see the following:

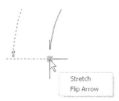

- Since arc length will not work with Continue and Baseline, only Stretch and Flip Arrow will appear.

8.24 EDITING DIMENSION BLOCKS USING THE RIGHT-CLICK MENU

- On the other hand, if you select a dimension block and right-click, you will see the following shortcut menu:

- You can change the dimension style used to insert the selected dimension block. Or better yet, you can save the changes you made as a new dimension style, as shown below:

- You can also change the Precision of the dimension text:

8.25 EDITING DIMENSION BLOCKS USING QUICK PROPERTIES AND PROPERTIES

- If you select a dimension block, Quick Properties will come up automatically. You will see something like the following:

- You have the ability to modify the dimension style and choose the Annotative feature (which will be discussed in the next chapter). You will also find the exact measurement of the selected dimension, and you can change it using the Text override field.
- Properties on the other hand will allow you to make global changes to the selected dimension blocks. Simply select the dimension block(s), right-click, and then select the Properties option (you can double-click as well). You will see something like the following:

- Using the Properties palette you can change everything related to the dimension block or the dimension style of the block (in the above example we show the variables you can control under the Fit category).

QUICK DIMENSIONS AND EDITING

 Practice 8-7

1. Start AutoCAD 2012.
2. Open **Practice 8-7.dwg**.
3. Input the following dimensions using all the techniques you have learned:

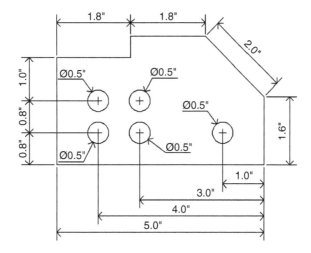

4. Using all the editing techniques make the following changes:

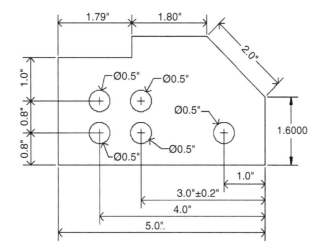

5. Save and close the file.

8.26 HOW TO CREATE A MULTILEADER STYLE

- A multileader is a replacement for the normal leaders that used to exist in AutoCAD. Leaders used to follow the current dimension style, and they were always single ones. A multileader has its own style, and a single leader can point to a different location in the drawing.
- A multileader style allows you to set the characteristics of a multileader block. To start this command, go to the **Annotate** tab, locate the **Leaders** panel, and then select the **Multileader Style** button:

- You will see the following dialog box:

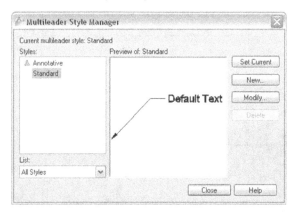

- As you can see there are two predefined styles: Standard (the default) and Annotative. Click the **New** button to create a new multileader style, and you will see the following dialog box:

- Input the name of the new style, and then click the **Continue** button. There are three tabs, and each one will control part of the multileader block:
  - Leader Format.
  - Leader Structure.
  - Content.

8.26.1 Leader Format Tab

- The following shows the **Leader Format** tab:

- Here, you can change all or any of the following:
  - Edit the **Type** of the leader, and choose one of three choices: Straight, Spline, or None. Here is an example of both the Straight and Spline options:

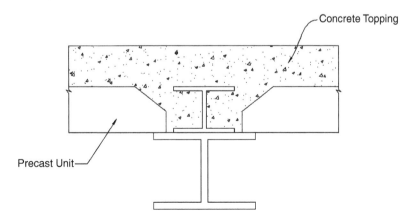

- Edit the **Color**, **Linetype**, and **Lineweight**.
- Select the **Arrowhead** shape and its size.
- Set the distance of the dimension break from any two blocks that intersect.

8.26.2 Leader Structure Tab

- The following shows the **Leader Structure** tab:

- You can change all or any of the following:
  - Specify if you want to change the **Maximum leader points**, and then input the desired value. By default, this value is 2; that is, the first point points to the geometry and the second point is the end of the multileader:

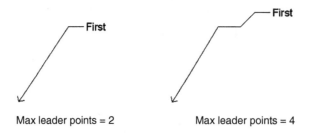

Max leader points = 2   Max leader points = 4

  - Specify whether you want to change **First segment angle** and **Second segment angle**. If yes, specify the angle values.
  - Specify whether you want AutoCAD to **Automatically include landing**. If yes, specify the **landing length.**
  - Specify whether the multileader will be **Annotative** (discussed in Chapter 9).

8.26.3 Content Tab

- The following shows the **Content tab**:

- In AutoCAD, there are two multileader types:
  - Mtext.
  - Block (either predefined or user-defined).
- The following illustration shows the two types:

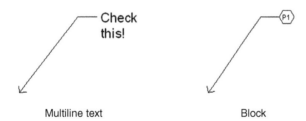

Multiline text   Block

- If you select the Mtext option you will be able to change all or any of the following:
  - If there is **Default text**.
  - Select **Text style**, **Text angle**, **Text color**, and **Text height** (if Text style's height = 0).
  - Select whether the text is **Always left justify** and with **Frame**.

- Choose whether the leader connection is horizontal or vertical. If vertical, then edit the position of the text relative to the landing for both left and right leader lines, and then control the gap distance between the end of the landing and the text:

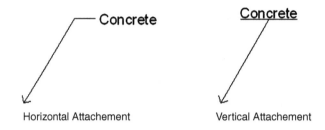

Horizontal Attachement   Vertical Attachement

- If you select the **Block** option, you can change all or any of the following:

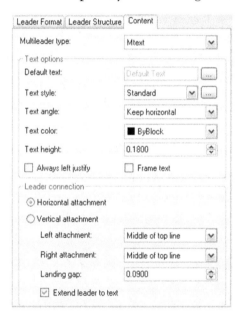

- Specify the **Source block**:

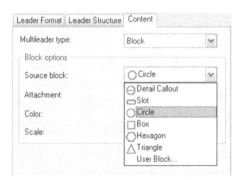

- Specify the **Attachment** position, **Color** of the attachment, and finally, the **Scale** of the attachment.

8.27 INSERTING A MULTILEADER DIMENSION

- This group of commands will allow you to add a single multileader, add a leader to an existing multileader, remove a leader from an existing multileader, and align and group an existing multileader. You will always start with the Multileader command, which will insert a single leader. To start this command, go to the **Annotate** tab, locate the **Leaders** panel, and then select the **Multileader** button:

- You will see the following prompts:

```
Specify leader arrowhead location or [leader Landing
first/Content first/Options] <Options>:
Specify leader landing location:
```

- First, specify the leader arrowhead location, then specify the leader landing location, and then type the text you want to appear beside the leader. To add a leader to an existing multileader, go to the **Annotate** tab, locate the **Leaders** panel, and then select the **Add Leader** button:

- You will see the following prompts:

```
Select a multileader:
1 found
Specify leader arrowhead location:
Specify leader arrowhead location:
```

- To remove a leader from an existing multileader, go to the **Annotate** tab, locate the **Leaders** panel, and then select the **Remove Leader** button:

- You will see the following prompts:

```
Select a multileader:
1 found
Specify leaders to remove:
Specify leaders to remove:
```

- To align a group of multileaders, go to the **Annotate** tab, locate the **Leaders** panel, and then select the **Align** button:

- You will see the following prompts:

```
Select multileaders: 1 found
Select multileaders: 1 found, 2 total
Select multileaders:
Current mode: Use current spacing
Select multileader to align to or [Options]:
Specify direction:
```

- To collect a group of similar multileaders to be a single leader, go to the **Annotate** tab, locate the **Leaders** panel, and then select the **Collect** button. This command works only with leaders containing blocks:

- You will see the following prompts:

```
Select multileaders:
Select multileaders:
Specify collected multileader location or
[Vertical/Horizontal/Wrap] <Horizontal>:
```

CREATING MULTILEADER STYLES AND INSERTING A MULTILEADER

 Practice 8-8

1. Start AutoCAD 2012.
2. Open **Practice 8-8.dwg**.
3. Create a new multileader style based on Standard, using the following information:
   a. Name = Texture and Painting
   b. Arrowhead symbol = Dot small
   c. Arrowhead size = 0.35
   d. First segment angle = 0
   e. Automatically include landing = Off
   f. Multileader type = Block
   g. Source block = Circle
4. Create a new multileader style based on Standard, using the following:
   a. Name = Material
   b. Leader format = Spline
   c. Arrowhead symbol = Right angle
   d. Arrowhead size = 0.25
5. Make layer Dimension current.

6. Using both styles insert the following multileaders:

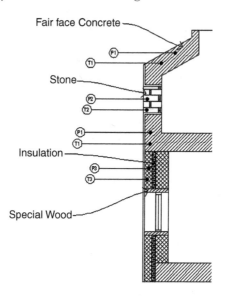

7. Using Add leader, Align, and Collect try to get the following:

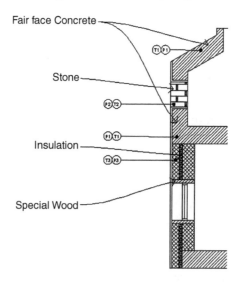

8. Save and close the file.

NOTES

CHAPTER REVIEW

1. The Arc length command should be used only with arcs and polylines, not with circles.
   a. True
   b. False
2. Symmetrical and Deviation are two types of _____ .
3. Continue and Baseline can't work with which of the following:
   a. Linear
   b. Radius
   c. Ordinate
   d. Angular
4. There are two types of multileader blocks.
   a. True
   b. False
5. When creating a new dimension style, which of the following statements is NOT correct:
   a. You can create a dimension style affecting all types of dimensions.
   b. You have to select the existing dimension style to start with.
   c. You can't create a dimension style affecting only one type of dimension.
   d. You can create a sub-style.
6. You can show _____ units and _____ units in a dimension block.
7. When using a multileader style there should be always a landing in your block.
   a. True
   b. False
8. Collect and Align are _____ commands.

CHAPTER REVIEW ANSWERS

1. a
3. b
5. c
7. b

Chapter 9   PLOTTING

In This Chapter

◇ The difference between Model space and Paper space
◇ How to create a new layout using different methods
◇ How to create and control viewports
◇ Plot style tables and creating and implementing them
◇ Using the Annotative feature
◇ Creating and viewing DWFs

9.1 WHAT IS MODEL SPACE AND PAPER SPACE?

- AutoCAD provides two spaces; one for creating your drawing, which is called Model space, and the other for plotting your drawing, which is called Paper space. There is only one Model space in each drawing; in contrast, there are an infinite number of Paper spaces per file, and each one is called a layout.
- Each layout is linked to a Page Setup where you specify everything related to plotting including the plotter, paper size, and paper orientation (Portrait or Landscape).
- After you create the layout and link it to a Page Setup, you will insert a title block, and then add viewports (which represent a portion of the Model space). Afterwards, you will set up the scale of the viewports.

9.2 INTRODUCTION TO LAYOUTS

- The layout is the place you will plot your drawing from. Each layout is linked to a **Page Setup**, **Objects**, such as the title block, text, dimensions, and finally, **Viewports**, which will be covered separately in the coming discussion. See the roadmap illustrated below:

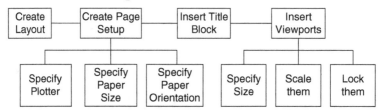

- Each layout should have a name. By default when you create a new drawing using the *acad.dwt* template, two layouts will come with it: Layout1 and Layout2. These two preset layouts are empty, so if you want to use them, take the following steps:
  - Rename the file to the desired name.
  - Create a Page Setup and link the layout to it, or you can simply link an existing Page Setup.
  - Insert a title block.
  - Insert viewports, scale them, and lock them.

9.3 CREATING A NEW LAYOUT FROM SCRATCH

- This method will allow you to create a new layout from scratch. It includes the following steps:
  - Right-click on any existing layout name (the tab at the lower-left corner of the screen and above the Command window), and a shortcut menu will appear; select the **New Layout** option:

- A new layout will be added with a temporary name. You should rename it by right-clicking and selecting the Rename option (you can click the temporary name, it will become editable, and you can input the new name):

- Link the new layout with a Page Setup. To do that, right-click the name of the layout, and then select Page Setup Manager:

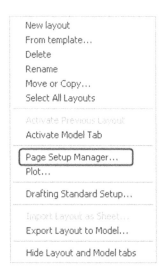

- The following dialog box will be shown:

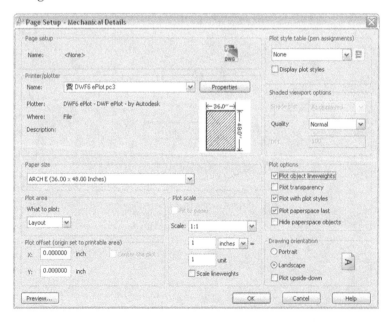

■ At the top of the dialog box, you will see the name of the **Current layout**, and on the bottom you will see the **Selected page setup details**, which is a summary of the current Page Setup. Finally, you will see a checkbox, **Display when creating a new layout**. This checkbox will force this dialog box to appear every time you go to a newly created layout.

■ By default, AutoCAD will create a Page Setup with the same name of the layout. Click the **Modify** button to modify it, and you will see the following dialog box:

- Select the desired **Plotter** (this plotter should be connected and configured).
- Select the desired **Paper size**.
- Specify **What to plot**. Always leave it as **Layout** (the other option is for Model space printing).
- Input **Plot offset**. If you are printing from layout, leave the values as zeros.
- Input **Plot scale**. If you are printing from layout then you will use viewports (this is our next topic), and each viewport will hold its own scale. Hence, you will set the layout plot scale to 1=1. Specify as well if you want to **Scale lineweights**.
- Select **Plot style table (pen assignments)** (this topic will be discussed at the end of this chapter).
- **Shaded viewport options** are for 3D modeling in AutoCAD.
- Leave **Plot options** as the default values.
- Select **Drawing orientation**, whether **Portrait** or **Landscape**.
- The plotter will print from top-to-bottom; select the checkbox if you want it to print otherwise.
- Click **OK**. The Page Setup you create will be available for all layouts in the current drawing file.
- You will be back to the first dialog box. Select the Page Setup and click **Set Current.** (You can also double-click the name of the Page Setup.) Now the current layout is linked to the Page Setup you selected.
- To modify the settings of an existing Page Setup click **Modify**. To import a saved Page Setup from an existing file click **Import**.

9.4 CREATING A NEW LAYOUT USING A TEMPLATE

- This procedure will allow you to import a layout from a template file, including the Page Setup and any other contents such as the title block, viewports, text, etc. To do this, take the following steps:
  - Right-click on any existing layout, and you will see the following menu; select the From template option:

- The normal Open file dialog box will pop up so can you can select the desired template file. Select the desired template, click Open, and you will see the following dialog box:

- Click one of the listed layouts and click OK. You will see the newly imported layout in your drawing.

9.5 CREATING A LAYOUT USING COPYING

- These two methods will allow you to create copies of an existing layout. The first method is similar to a method used in Microsoft Excel. Take the following steps:
  - Select the desired layout.
  - Hold the [Ctrl] key on the keyboard and hold and drag the mouse to the new position of the newly copied layout:

ınd: *Cancel*

- • Rename the new layout.
- ▪ Another way to copy layouts is to select the desired layout and then right-click and a shortcut menu will appear. Select the Move or Copy option:

- ▪ You will see the following dialog box:

- ▪ A list of the current layouts is displayed. Select one of them and click the checkbox **Create a copy** on. A copy will be created, and you can rename it and make the necessary changes.
- ▪ Using the same option you can move a layout from its current position to the left or to the right. Alternatively, you can move a layout without this command by clicking the layout name, holding, and dragging to the desired location.

- So far, you have accomplished the following steps:
  - Creating a New Layout.
  - Creating a Page Setup.
  - Linking a Page Setup to the layout.
- Accordingly, when you select the newly created layout, you will see something like the following:

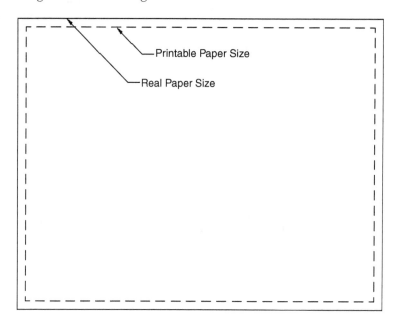

- The outer frame is the real paper size, and the inside frame (the dashed line) is the printable paper size, which is the paper size minus the printer's margins. Based on this layout, you will see exactly what will be printed and what will not, because anything outside the dashed line will not be printed. This proves that printing from layouts is WYSIWYG (What You See Is What You Get).
- You can bring in any layout from any DWG file using Design Center just like you would bring in blocks, layers, dimension styles, text styles, etc.

CREATING NEW LAYOUTS

Practice 9-1

1. Start AutoCAD 2012.
2. Open **Practice 9-1.dwg**.

3. Select Layout 1, and delete the existing viewport.
4. Click then right-click Layout 1, and select Page Setup Manager.
5. Create a new Page Setup using the following information:
   a. Name = Architectural Plan A3
   b. Plotter = DWF6 ePlot.pc3
   c. Paper size = ISO A3 (420.00 x 297.00 MM)
   d. What to plot = Layout
   e. Plot scale = 1:1
   f. Drawing orientation = Landscape
   g. Click OK to end the creation process
6. Make sure you are selecting the newly created Page Setup, and click **Set Current** to link this Page Setup to the current layout.
7. Rename the layout Architectural Plan (A3 Size).
8. Make layer Title Block the current layer.
9. Insert the file called A3 Size Title Block.dwg to be your title block, using the insertion point 0,0.
10. Create a copy of the newly created layout, and name it Mechanical.
11. Delete Layout 2.
12. Right-click any existing layout, and select the From template option.
13. Import the ISO A1 Layout from Tutorial-mArch.dwt.
14. Move the ISO A1 Layout to be the first layout after Model tab.
15 Save and close the file.

9.6 CREATING VIEWPORTS

- When you visit a layout for the first time (just after creation) you will see that a single viewport appears at the center of the paper. A *viewport* is a window of any shape containing the view of your Model space, initially scaled to the size of the window.
- Viewports inserted in layouts can be tiled, will be scaled, and can be printed.
- You can add viewports to layouts by:
  - Adding a single rectangular viewport.
  - Adding multiple rectangular viewports.
  - Adding a single polygonal viewport.
  - Converting an object to a viewport.
  - Clipping an existing viewport.

9.6.1 Adding Single Rectangular Viewports

- This command will allow you to add single rectangular viewports to a layout. You should specify two opposite corners for the area of the viewport. To issue this command, go to the **View** tab, locate the **Viewports** panel, and then select the **Rectangular** button:

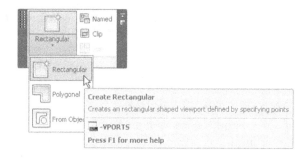

- The following prompts will be shown:

```
Specify corner of viewport or [ON/OFF/Fit/Shadeplot
/Lock/Object/Polygonal/Restore/Layer/2/3/4] <Fit>:
Specify opposite corner:
```

- Specify the two opposite corners to create a single rectangular viewport.
- This is what you will get:

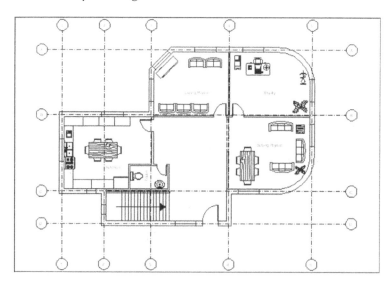

9.6.2 Adding Multiple Rectangular Viewports

- This command will allow you to add multiple rectangular viewports to a layout. You should specify two opposite corners for the area of the viewports.

To issue this command, at the Command window type **+vports**, then press [Enter]. Press [Enter] again to accept the default value for the next prompt, and you will see the following dialog box:

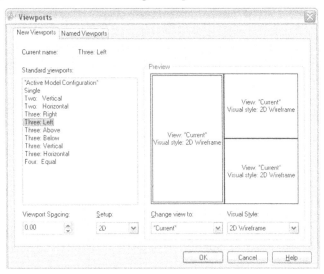

- Select the desired display from the list. By default, the **Viewport Spacing** value = 0 (zero), which means the viewports will be tiled. If you want them separated input a value greater than 0 (zero). Click **OK**, and you will see the following prompts:

```
Specify first corner or [Fit] <Fit>:
Specify opposite corner:
```

- See the following illustration:

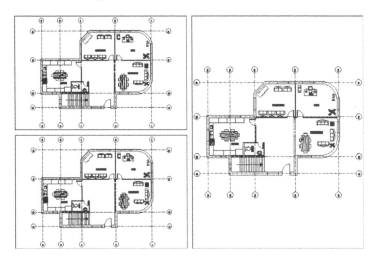

9.6.3 Adding a Polygonal Viewport

- This command will allow you to add a polygonal viewport consisting of both straight lines and arcs. To start this command, go to the **View** tab, locate the **Viewports** panel, and then select the **Polygonal** button:

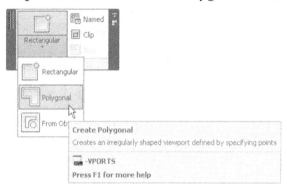

- You will see the following prompts:

```
Specify start point:
Specify next point or [Arc/Length/Undo]:
Specify next point or [Arc/Close/Length/Undo]:
```

- These prompts look like the Polyline command prompts.
- See the following illustration:

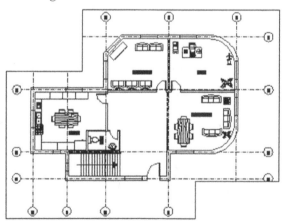

9.6.4 Creating Viewports by Converting Existing Objects

- This command will allow you to convert any existing object (should be a single object, such as a polyline or circle; lines and arcs are not allowed) to the viewport. To start this command, go to the **View** tab, locate the **Viewports** panel, and then select the **From Object** button:

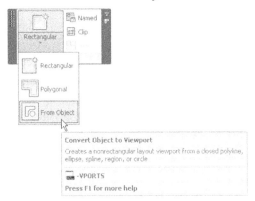

- You will see the following prompt:

```
Select object to clip viewport:
```

- See the illustration below:

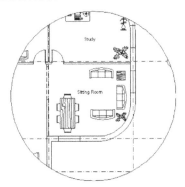

9.6.5 Creating Viewports by Clipping Existing Viewports

- This command will allow you to clip an existing viewport and create a new shape. To start this command, go to the **View** tab, locate the **Viewports** panel, and then select the **Clip** button:

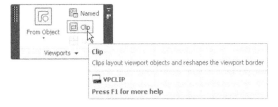

- You will see the following prompts:

```
Select viewport to clip:
Select clipping object or [Polygonal] <Polygonal>:
Specify start point:
Specify next point or [Arc/Length/Undo]:
Specify next point or [Arc/Close/Length/Undo]:
```

- First, select an existing viewport. You can select a polyline drawn previously, or you can draw any irregular shape using the **Polygonal** option (which is identical to the Polygonal viewport command discussed above). See the illustration below:

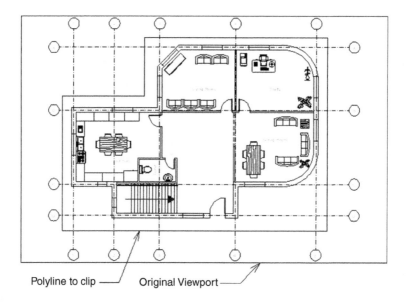

Polyline to clip ————/   Original Viewport ————

9.6.6 Working with Viewports After Creation

- There are two ways to work with viewports:
  - Outside the viewport, which means you will deal with it like any other object in your drawing. You can select the viewport from its frame, and erase, copy, move, scale, stretch, rotate, etc.

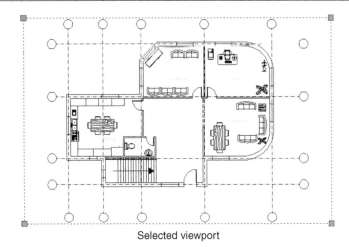

Selected viewport

- The second way is from inside the viewport, which you can achieve by double-clicking inside the viewport. This mode will allow you to zoom, pan, scale, etc., the objects inside the viewport. To return to the first mode, double-click outside the viewport:

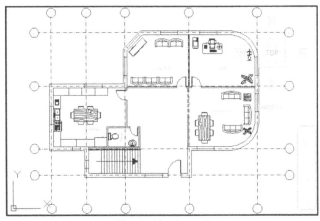

User inside viewport

9.7 SCALING AND MAXIMIZING VIEWPORTS

- By default when you insert a viewport, AutoCAD will zoom the whole drawing into the area you specified as the viewport size. This viewport is *not-to-scale*, and you should set the scale relative to the Model space units. Take the following steps:

- Double-click inside the desired viewport (or you can select only the viewport's frame).
- Look at the right side of the status bar, and you will see the Viewport Scale list as shown below:

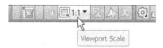

■ Click the list that contains all scales, something like the following:

■ Select the desired scale for your viewport. You can also select the **Custom** option, and the following dialog box will be displayed:

■ To add a new scale, select the **Add** button, and the following dialog box will be displayed:

- Input the desired scale, and then click **OK** twice.
- After the scaling process is finished, either the scale will be perfect for the area of the viewport, or it will be either too big or too small. The solution for such a problem is to change the size of the viewport or to change the scale.
- After setting the scale, you can use the Pan command but not the Zoom command because this will ruin the scale you set. To avoid this problem, you can lock the display of the viewport by the selecting viewport(s) and then clicking the golden opened lock in the status bar (you have to be inside the viewport or have the border of the viewport selected), the golden lock will change to a blue lock, and the viewports will be locked accordingly:

- The Maximize function will allow you to maximize the viewport to fit the size of the screen temporarily. This will give you the needed space to edit objects as you wish without leaving the layout and going to the Model space. Go to the status bar and click the **Maximize Viewport** button as shown:

- The same button will restore the original size of the viewport.

9.8 FREEZING LAYERS IN A VIEWPORT

- The freezing function will freeze layer(s) in Model space and all viewports in all layouts. If you want to freeze a layer in a certain viewport, take the following steps:
  - Double-click the desired viewport.
  - Go to the **Home** tab, locate the **Layers** panel, select the layer list, and click the icon **Freeze or thaw in current viewport** for the desired layer. See the illustration below:

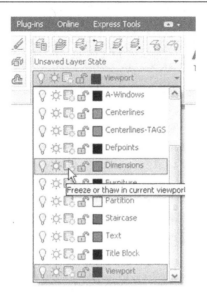

- You also have the ability to freeze/thaw the layer in all viewports except the current viewport. To do this start the Layer Properties Manager, select the desired layer(s), right-click and select **VP Freeze layer**, and then the **In All Viewports Except Current** option, as shown below:

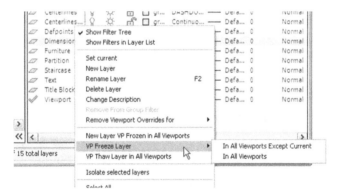

9.9 LAYER OVERRIDE IN A VIEWPORT

- In all viewports a layer will be displayed with the same color, linetype, line-weight, and Plot Style, but you can change these settings in one viewport. This is called layer override. Take the following steps:
  - Double-click inside the desired viewport.
  - Issue the Layer Properties Manager command.

- Under VP Color, VP Linetype, VP Lineweight, or VP Plot Style make the desired changes.
- These changes will only affect the current viewport.
■ See the following illustration:

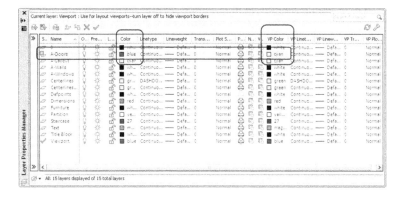

- In the above example layer **A-Doors** has the color blue (applies in Model space and all other viewports) and an override color of cyan in the current viewport.
- Also, notice that the layer **A-Doors** row is shaded with a different color.

CREATING AND CONTROLLING VIEWPORTS

 Practice 9-2

1. Start AutoCAD 2012.
2. Open **Practice 9-2.dwg**.
3. Make layer Viewport the current layer.
4. Switch to the D-Size Architectural Plan layout.
5. Insert a single viewport to fill the entire area.
6. Set the scale to ½″ = 1′, and lock the viewport.
7. Switch to the D-Size Arch Details layout.
8. Insert viewports using Three: Right, with Viewport Spacing = 0.25, and select an area to fill the paper size.
9. Select the borders of the big viewport at the right, and scale it to ¼″ = 1′.
10. Using the grips, shrink the area of the viewport to something suitable for the scale chosen.
11. Select the two viewports at the left, and set the scale to ¾″ = 1′.

12. For the top viewport and using the Pan command, set the view to show the Kitchen and Toilet.
13. For the bottom viewport and using the Pan command, set the view to show the Living Room.
14. Select the three viewports and lock them.
15. Switch to the ANSI B Size Architectural Plan layout.
16. Draw a big circle at the center of the paper using Create from the Object command, and convert the circle to a viewport.
17. Double-click inside the new viewport, set the scale to ½″ = 1′, and then pan to show the Study.
18. In the current viewport, thaw Centerline and Centerline-TAGS.
19. Switch to the other layouts to make sure that these two layers were frozen only in this viewport.
20. Change the color of layer A-Walls in this viewport to red.
21. Save and close the file.

9.10 PLOT STYLE TABLES – FIRST LOOK

- Plot styles are used to convert the colors used in the drawing to printed colors. The default setting is to keep the same color as the printer. Before AutoCAD 2000, there was only one type of conversion method, but more recent versions have a new feature called Plot Style Tables. There are two types of Plot Style Tables:
  - Color-dependent Plot Style Table
  - Named Plot Style Table

9.11 COLOR-DEPENDENT PLOT STYLE TABLE

- This Plot Style table is a simulation of the only method that existed before AutoCAD 2000. The essence of this method is simple: for each color used in your drawing you will specify the color to be used at the printer. AutoCAD will allow you to set the lineweight, linetype, etc., for each color. The problem with this method is that you only have 255 colors to choose from.
- Each time you create a Color-dependent Plot Style Table, AutoCAD will create a file with the extension *.ctb*. You can create Plot Style Tables from outside AutoCAD using the Control Panel in Windows, or from inside AutoCAD using the menu bar.

- From outside AutoCAD, start the Control Panel in Windows, double-click the **Autodesk Plot Style Manager** icon, or show the menu bar, and then select **Tools/Wizards/Add Plot Style Table**. You will see the following dialog box:

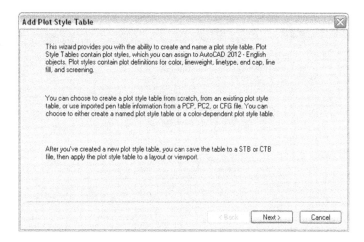

- The first screen is an introduction to the whole concept of Plot Styles; read it to understand the next steps. When done, click the **Next** button, and you will see the following dialog box:

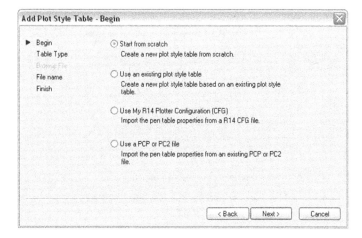

- AutoCAD will list four possible choices:
  - Start from scratch.

- Use an existing Plot Style.
- Import an AutoCAD R14 CFG file, and create a Plot Style from it.
- Import a PCP or PC2 file, and create a Plot Style from it.

■ Select **Start from scratch**, click the **Next** button, and you will see the following dialog box:

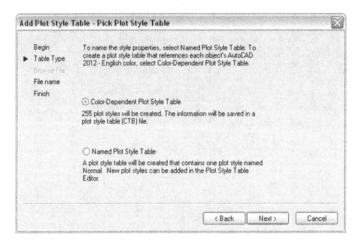

■ Select **Color-Dependent Plot Style Table**, click the **Next** button, and you will see the following dialog box:

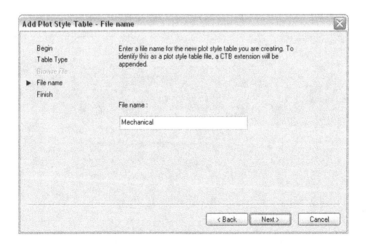

■ Input the name of the new Plot Style, click the **Next** button, and you will see the following dialog box:

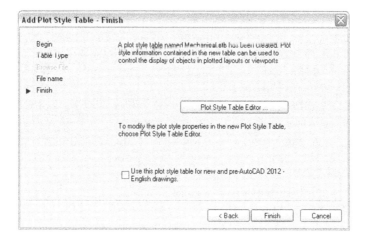

- Select the **Plot Style Table Editor** button, and the following dialog box will be displayed:

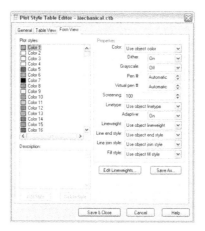

- From the left part, select the color you used in your drawing file, and then at the right, change all or any of the following settings:
  - Change **Color** to be used in the plotter.
  - Switch **Dither** on/off. This option will be dimmed if your printer or plotter doesn't support dithering. Dither is a method used to give the impression of using more colors than the 255 colors used by AutoCAD. It is better to leave this option off, but it should be on if you want **Screening** to work.
  - Turn **Grayscale** on/off. This option is good for laser printers.
  - Change **Pen #**. This option is valid for the old types of plotter – pen plotters – which are not used these days.

- Change **Virtual pen #**. Used for non-pen plotters to simulate pen plotters by assigning a virtual pen for each color; leave it on **Automatic**.
- Change **Screening**. This option will reduce the intensity of the shading and fill the hatches, hence reducing the amount of ink used. This option depends on **Dither**.
- Change **Linetype**. Set a different linetype for the color or leave it as the object's linetype.
- Change **Adaptive**. This option will change the linetype scale of all objects using the color to start a segment and end a segment. Turn this option off if the linetype scale is important for your drawing.
- Change **Lineweight**. This option will change the lineweight for the color selected.
- Change **Line end style**. This option will allow you to select the end style for lines; choose one of the following: Butt, Square, Round, and Diamond.
- Change **Line join style**. To select the line join shape, choose from Miter, Bevel, Round, and Diamond.
- Change **Fill style**. This option will set the fill style for the filled area in the drawing (good for trial printing).

- Click **Save & Close**. Then click **Finish.**
- Your last step should be linking your Plot Style with a layout. Take the following steps:
  - Select the desired layout, then start the **Page Setup Manager**.
  - Select the name of the current Page Setup and click **Modify**.
  - At the upper-right part of the dialog box, under **Plot style table (pen assignments),** select the desired Plot Style Table:

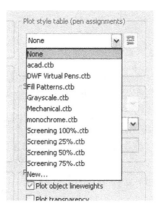

  - Click the Display Plot Styles checkbox on.

- For each layout, you can assign one .ctb file. In order to see the linetype and lineweight of the objects you have to click the **Show/Hide Lineweight** button on the status bar on, as shown below:

9.12 NAMED PLOT STYLE TABLE

- This method does not depend on colors. The created Plot Style Tables will be linked later on with the layers, so you may have two layers with the same color, yet they will print with different colors, linetypes, and lineweights.
- A Named Plot Style Table has a file extension of *. *stb*. The creation procedure of a Named Plot Style is identical to creating a Color-dependent Plot Style, except the last step, which is configuring the **Plot Style Table Editor**. You can create it from outside AutoCAD using the Control Panel and double-clicking the **Autodesk Plot Style Manager** icon, or you can show the menu bar, and then select **Tools/Wizards/Add Plot Style Table**. You will see the same screen you saw while creating a Color-dependent Plot Style Table.
- Follow the same steps until you reach the **Plot Style Table Editor** button. Click it and you will see the following dialog box; click the **Add Style** button:

- As you can see you can change all or any of the following:
  - Input the Name of the style and a brief Description.

- Change the Color to be used in the plotter.
- We explained the rest of features in the previous section on Color-dependent Plot Style Tables.

- You can add as many styles as you wish in the same Named Plot Style. Click **Save & Close**. Then click **Finish**.

- Linking a Named Plot Style Table with any drawing is a bit more complicated than linking a Color-dependent Plot Style Table:
  - The first step is a precautionary step. You may want to print a drawing, but discover you can only use .ctb files. To solve this problem you have to convert one of the .ctb files to a .stb file. At the Command window type **convertctb**, and a dialog box listing all the .ctb files will appear. Select one of them, keeping the same name, or give it a new name, and then click **OK**. You will see the following dialog box:

  - Convert the drawing from a Color-dependent Plot Style to a Named Plot Style. At the Command window type **convertpstyles**, and you will see the following warning message:

  - Click **OK**, and you will see the following dialog box:

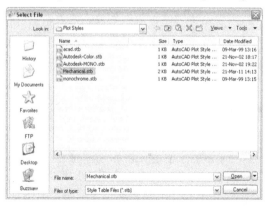

- Select the newly created Named Plot Style Table, click **Open**, and you will see the following prompt:

```
Drawing converted from Color Dependent mode to Named
plot style mode.
```

- The second step is to select the desired layout. Start the **Page Setup Manager**, and at the upper-right part of the dialog box, under **Plot style table (pen assignments),** select the name of the newly created named Plot Style Table. Click the **Display plot styles** checkbox on, and end the Page Setup Manager command:

- Select the **Layer Properties Manager**, and for the desired layer(s), click the name of the Plot Style in the **Plot Style** column:

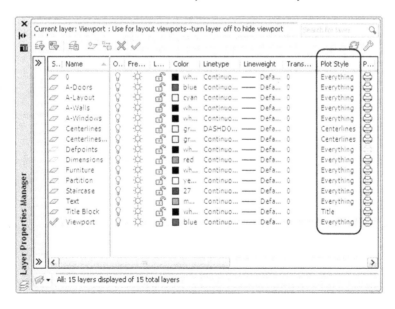

- You will see the following dialog box:

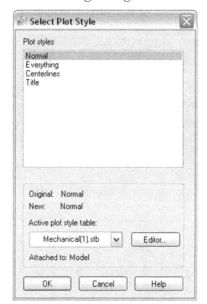

- Select the desired Plot Style. When done, click **OK**.
- Repeat the same steps for the other layers.
- You may need to type at the Command window **regenall** (which means regenerate all viewports) in order to see the effects of your changes.

 ▪ If you create a new drawing using *acad.dwt* your drawing will only accept a Color-dependent Plot Style Table. Use *acad-Named Plot Styles.dwt* to create a new drawing that will only accept a Named Plot Style Table.

COLOR-DEPENDENT PLOT STYLE TABLES

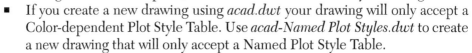 **Practice 9-3**

1. Start AutoCAD 2012.
2. Open **Practice 9-3.dwg**.
3. Create a new Color-dependent Plot Style Table and call it Architectural using the following settings:

Color	1	2	3	4	5	6	7	27
Plot color	Black	Black	Use object color	Black	Black	Black	Black	Black
Lineweight	0.3	0.3	0.3	0.3	0.7	0.3	0.3	0.3

4. Switch to the D-Size Architectural Plan layout.
5. Link this layout to the Architectural Plot Style.
6. Check that the Show/Hide Lineweight button on the status bar is turned on.
7. Compare the doors with the other objects, and you will find they are thicker, because color (5) was used for both title block and doors with a 0.7 lineweight.
8. Save and close the file.

NAMED PLOT STYLE TABLES

 Practice 9-4

1. Start AutoCAD 2012.
2. Open **Practice 9-4.dwg**.
3. Create a new Named Plot Style Table and name it Architectural using the following settings:

Style name	Color	Lineweight
Everything	Black	0.3
Centerlines	Green	0.5
Title	Black	0.7

4. Link the newly created table with the D-Size Architectural Plan layout.
5. Using the Layer Properties Manager make the following changes:
   a. Layer = Centerlines and Centerlines-TAGS linked to Centerlines
   b. Title Block linked to Title
   c. The rest of layers linked to Everything
6. Zoom in to compare the Centerlines lineweight to the other objects.
7. Save and close.

9.13 PLOT COMMAND

- This is our final step. This command sends whatever we set in the layout to the specified plotter. To issue this command, go to the **Output** tab, locate the **Plot** panel, and then select the **Plot** icon:

- You will see the following dialog box:

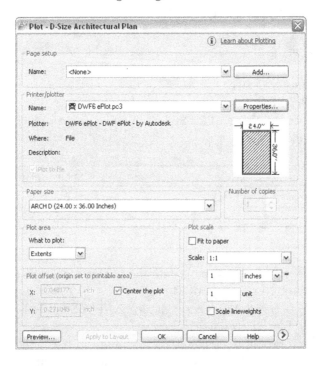

- This dialog box is identical to the Page Setup settings. If you modify any of these settings, AutoCAD will separate the Page Setup from the current layout. To send the layout to the plotter, click **OK**.
- To save the settings of this dialog box with the layout, select the **Apply to Layout** button.
- Click the **Preview** button to preview the final printed drawing on the screen before it is printed.
- You can also preview your drawing from outside this dialog box by going to the **Output** tab, locating the **Plot** panel, and then selecting the **Preview** button:

9.14 WHAT IS THE ANNOTATIVE FEATURE?

- Since we will always print from layout, we need to use viewports. And since we will use viewports we have to set the viewport scale for each viewport. The viewport scale will affect all objects in the Model space. So, if you hatch, type text, insert dimensions, insert a multileader, or insert a block that contains text in Model space, all of these objects will be scaled. If they are scaled down, then the text and dimensions will be unreadable, and the hatch will look like solid hatching.
- What we need is a feature that will scale everything except the annotation objects (hatches, text, and dimensions), and this feature is the **Annotative** feature.
- The Annotative feature can be found in different places:
  - You will find it in Text style, under Size:

  - You will find it in Dimension style; go to the Fit tab:

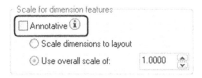

  - You will find it in Multileader style; go to the Leader Structure tab:

- • You will find it in the Hatch context tab, in the Options panel:

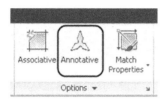

- • You will find it in the Block creation command, under Behavior:

■ How will you know that you are dealing with a style that supports the Annotative feature? You will see a special symbol by its name. See the illustration below:

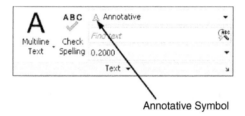

Annotative Symbol

■ How will you know that an object was inserted using a style or a command that supports the Annotative feature? Simply hover over it, and if you see something like the following, you know it was inserted with the Annotative feature on:

- You will take the following steps to use the Annotative feature:
    - Create your drawing in Model Space, without any annotation objects (hatches, text, dimensions, multileaders, and blocks with text).
    - Select the desired layout, add viewports, and then scale them. This scale will be for both the annotation objects and the viewport.
    - Double-click inside the viewport to make it active.
    - Insert all the annotation objects you need.
    - Once you insert annotation objects in the viewport, you can see them in this viewport and any other viewport holding the same scale value.
    - Changing the scale of the viewport will result in losing the annotation object.
    - To show the annotation object in more than one viewport with different scale values, right-click on the **Annotation Visibility** button (at the right portion of the status bar), and you will see the following menu:

    - Select either to Show Annotation Objects for Current Scale Only, or Show Annotation Objects for All Scales.
    - Control how to add scale values to the viewport: Automatic or Manual. Right-click on the **Add Scale** button (at the right portion of the status bar), and you will see the following menu:

    - If you use a zooming command inside the viewport, this will ruin the current viewport scale. To get it back click the **Synchronize** button (at the right portion of the status bar), and you will see the following message (use the lock in the status bar to lock the viewport scale to avoid this problem):

- Clicking this message will restore the viewport scale.
- To make an annotation object appear in viewports with different scales, select it, right-click, and select the **Annotative Object Scale** option. You will see the following sub-menu:

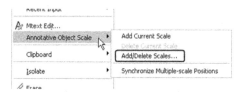

- Choose the **Add/Delete Scales** option, and the following dialog box will appear:

- Select the **Add** button, and the following dialog box will appear:

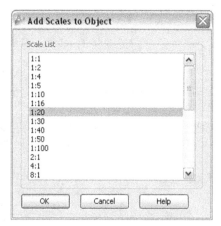

- Select the desired scale value, and click **OK** twice.

ANNOTATIVE FEATURE

 Practice 9-5

1. Start AutoCAD 2012.
2. Open **Practice 9-5.dwg**.
3. Make layer Dimensions current.
4. Switch to the Details layout.
5. There are three viewports; the first one at the left is scaled to 1:20, and the other two at the right are scaled to 1:10.
6. Go to the Annotate tab, locate the Dimensions panel, and check the available dimension styles. You will see only one with a distinctive symbol at its left, which is mechanical. This dimension style is annotative.
7. Double-click inside the big viewport.
8. Start the Radius command, select the big magenta circle, and then add the dimension block. Make sure it didn't appear in the upper-right viewport.
9. Add another Radius block to the small magenta circle.
10. Add two linear dimensions for the total width and the total height.
11. While you are still inside the same viewport, make layer Text current.
12. Make the text style Annotative current.
13. Using multiline text beneath the shape add the word Bearing.
14. Make layer Dimensions current again.
15. Click inside the upper-right viewport to make it current, and add a Radius dimension to one of the small circles.
16. Make layer Hatch current.
17. Click inside the lower-right viewport.
18. Start the Hatch command, make sure that Annotative is on, and set the scale = 10. Hatch the area between the two dashed lines, and then finish the Hatch command.
19. Using Add/Delete scales add scale 1:20 so the two hatches will appear in the big viewport.
20. You should have the following:

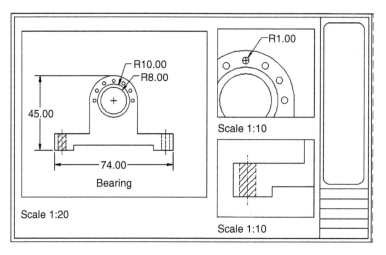

21. Save and close the file.

9.15 DESIGN WEB FORMAT (DWF) FILES

- Sharing files and data is becoming vital for everyone, and AutoCAD files are no exception. But, if you send .dwg files to others (joint venture companies, third parties, clients, vendors, etc.) you are taking the risk that your design could be stolen. Moreover, .dwg files are always bulky (the file size may reach to more than 50-100 MB depending on if it was 2D or 3D) and are very difficult to send via e-mail. Additionally, users need AutoCAD to open these files and not everyone has it. To solve these problems, AutoCAD allows us to plot to the **D**esign **W**eb **F**ormat (DWF) file type. A DWF file can't be modified, so you don't have to worry about your design being stolen, and the DWF size is relatively small compared to the DWG file size. To view a DWF you will need a free program called Autodesk Design Review, which comes on the same DVD as AutoCAD, or it can be installed from the Autodesk website. This software is for viewing and printing, but you can also use it to measure and red-line DWF files as well.
- Another version of DWF, called DWFx, can be viewed with both Windows Vista and Windows 7 using the Internet Explorer browser.

9.16 EXPORTING DWF, DWFX, AND PDF FILES

- This command will allow you to export your current DWG file to DWF, DWFx, and PDF files. To start this group of commands, go to the **Output** tab, locate the **Export to DWF/PDF** panel, and then select the **Export** button:

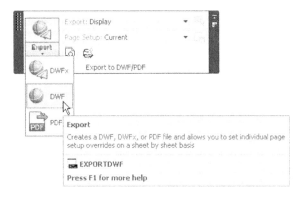

- You will see the following dialog box:

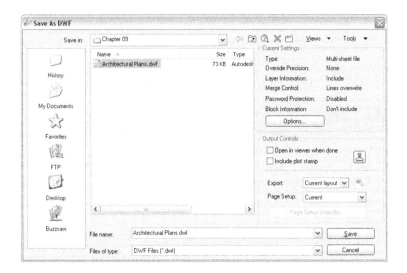

- At the top-right part of the dialog box, under **Current Settings**, AutoCAD lists the current settings, as shown below:

- To edit these settings, click the **Options** button, and the following dialog box will appear:

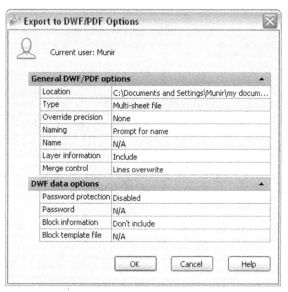

- Change all or any of the following settings:
  - Indicate the **Location** on the hard disk where you will save the DWF file.
  - File **Type** by default it is always Multi-sheet.
  - Indicate the **Override** precision; you can pick from Manufacturing, Architectural, etc. This setting will allow you to pick the dpi (dots per inch) precision, which will allow DWF users to measure distances accurately.
  - Input the **Name** of the DWF file, or let AutoCAD prompt you after executing the command.
  - Indicate whether to include **Layer** information. This will allow you to turn off layers in the Autodesk Design Review software.
  - Indicate whether to include a **Password**.
  - To finish, click **OK**.
- Under **Output Controls** you will see the following:

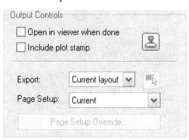

- Set all or any of the following:
  - To open a DWF using Autodesk Design Review automatically after executing the command.
  - To include a plot stamp.
  - Indicate what to export. If you execute the command from Model Space, then select Display, Extents, or Window. On the other hand, if you are exporting while you are at one of the layouts, then the available selections will be Current Layout or All layouts.
  - Select the Page Setup.
  - To finish, click the **Save** button.

9.17 USING THE BATCH PLOT COMMAND

- This command will allow you to produce DWF files containing multiple layouts from the current drawing and from other drawings. To issue this command go to the **Output** tab, locate the **Plot** panel, and then select the **Batch Plot** button:

- You will see the following dialog box:

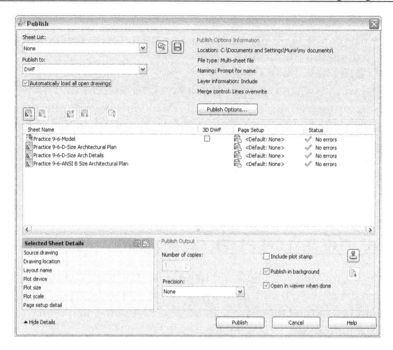

- You will see a table that contains Model Space and layouts. Set all or any of the following:
  - Indicate Publish to either the printer/plotter defined in the layout or DWF, DWFx, PDF.
  - Indicate whether to load all open drawings automatically.
  - Using the five buttons above the table, you can add a sheet, remove a sheet, and move any sheet up or down to specify its order relative to the document sheets. Finally, you can preview the sheet:

  - You can rename the sheets by clicking on any sheet name, and the name will become editable.
  - Click **Publish Options**, and the following dialog box will appear, which is identical to what we discussed in the Exporting section:

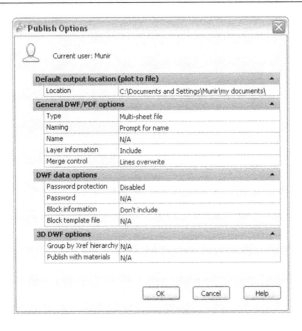

- When done click **OK**.
- Specify the number of copies.
- Select whether to include a plot stamp.
- Indicate whether you want to publish a DWF in the background.
- Indicate whether to open the DWF using Autodesk Design Review.
- Set the precision for the DWF file (dpi precision) for measurements in Autodesk Design Review.
  - To finish, click the **Publish** button. A final message will come up as a bubble similar to the following:

9.18 VIEWING DWF AND DWFX FILES

  - AutoCAD comes with Autodesk Design Review. You will see a shortcut on your Desktop after you install AutoCAD. Once you locate it, start it, and open a DWF or DWFx file (you can only open one file at a time), you will see the following:

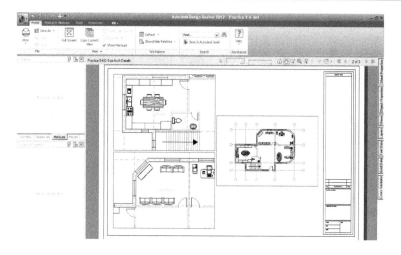

CREATING AND VIEWING A DWF FILE

 Practice 9-6

1. Start AutoCAD 2012.
2. Open **Practice 9-6.dwg**.
3. Produce a multi-sheet DWF file including all layouts except the Model Space sheet. Make sure to include layers in the DWF file. Save the file as Architectural Plans.dwf. Open it using Autodesk Design Review.
4. If you have Windows Vista or Windows 7, produce a DWFx file, and open it using Internet Explorer.
5. Save and close the file.

NOTES:

CHAPTER REVIEW

1. Which of the following is true about creating a new layout?
   a. You can bring in a layout from an existing template.
   b. You can create a new layout using the right-click menu at the name of an existing layout.
   c. You can bring in a layout using Design Center.
   d. All of the above.
2. Objects to be converted to a viewport should be _____ objects such as a circle or a polyline.
3. Using the Pan command after setting the scale of a viewport will ruin the scaling process.
   a. True
   b. Flase
4. You can freeze a layer in a certain viewport. You can change the color of a layer in a certain viewport.
   a. The first statement is correct, but the second is incorrect.
   b. Both statements are correct.
   c. Both statements are incorrect.
   d. The first statement is incorrect, but the second is correct.
5. Color-dependent Plot Styles are similar to what was used before AutoCAD 2000.
   a. True
   b. False
6. The best practice is to insert dimensions, text, and hatches in _____, using the _____ feature.
7. A _____ is better than a Color-dependent Plot Style Table.
8. There is no difference between DWF and DWFx files.
   a. True
   b. False
9 In order to insert an annotative object you have to be inside the viewport.
   a. True
   b. False

CHAPTER REVIEW ANSWERS

1. d
3. b
5. a
7. Named Plot Style Table
9. a

Chapter **10** **PROJECTS**

In This Chapter

◇ How to set up your drawing
◇ Architectural project in metric and imperial units
◇ Mechanical project – I in both metric and imperial units
◇ Mechanical project – II in both metric and imperial units

10.1 HOW TO PREPARE YOUR DRAWING FOR A NEW PROJECT

- In order to prepare your drawing for a new project, take the following steps:
  - Start a new drawing based on *acad.dwt* or *acadiso.dwt* if you want to use a color-dependent plot style table.
  - Start a new drawing based on *acad-Named Plot Styles.dwt*, or *acadISO-Named Plot Styles.dwt* if you want to use a named plot style table.
  - Set up the Drawing Units. From the Application Menu select **Drawing Utilities/Units**:

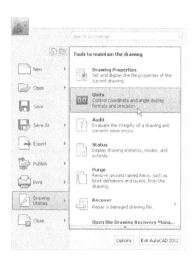

- You will see the following dialog box:

- Choose the desired **Length Type**. Select one of the following: Architectural (example: 2'-4 8/16"), Decimal (example: 25.5697), Engineering (example: 3'-5.6688"), Fractional (example: 16 3/16), Scientific (example: 8.9643E+03).
- Choose the desired **Angle Type**. Select one of the following: Decimal Degrees (example: 36.7), Deg/Min/Sec (example: 47d25'31"), Grads (example: 60.7g), Radians (example: 0.6r), Surveyor's Units (example: N 51d25'31" E).
- Set up the **Precision** for both length and angle units. For instance, choose the number of decimals for decimal units between zero and eight decimal places.
- The default AutoCAD setting for angles is Counter Clock Wise, but you can switch to **Clockwise**.
- Under **Insertion scale**, specify **Units to scale inserted content** (this was discussed in Chapter 6).
- Select the **Direction** button, and you will see the following dialog box:

- Change the East to a 0 (zero) angle, and the other angles will change as well (we recommend this setting be left as is). This step will end the Units command.
- Set up the drawing limits. The drawing limits contain your working area. You will specify the limits using two opposite corners; the lower-left corner and the upper-right corner. To set up the drawing limits correctly, answer two questions: what is the longest dimension in my drawing in both X and Y? And what is my AutoCAD unit (meter, centimeter, inch, or foot)?
- If the menu bar is showing, choose **Format/Drawing Limits**, or you can type **limits** at the Command window. You will see the following prompts:

```
Specify lower left corner or [ON/OFF] <0,0>:
Specify upper right corner <12,9>:
```

- Specify the lower-left corner and upper-right corner by typing or by clicking. To keep yourself from using an area outside these limits, AutoCAD gave you the ability to turn these areas on or off.
- Create layers.
- Start drafting.

10.2 ARCHITECTURAL PROJECT (IMPERIAL)

- Take the following steps:
    1. Start a new project using **acad-Named Plot Styles.dwt**.
    2. Switch off the grid.
    3. Set the units to the following:
        a. Architectural
        b. Precision 0'- ½"
        c. Units to scale inserted contents = inch
    4. Set up the drawing limits to be:
        a. Lower-left corner = 0,0
        b. Upper-right corner = 50',50'
    5. Double-click the mouse wheel in order to see the new limits.
    6. Create the following layers:

Layer Name	Layer Color
A-Door	Blue
A-Wall	White
A-Window	White
Dimension	Red
Furniture	White
Staircase	Magenta
Text	Green
Title Block	White
Viewport	9
Hatch	8

7. Save your file in the Chapter 10\Imperial folder and name it Ground Floor. dwg.
8. Make layer A-Wall current.
9. Draw the following architectural plan and the partitions inside using the following guidelines:
    a. Draw the outer shape using a polyline.
    b. Offset it to the inside using 6″ as distance.
    c. Explode the two polylines.
    d. Use the outer wall to draw the inner walls using all the commands you learned in this book. The inner wall is 4″.
10. This is the architectural plan:

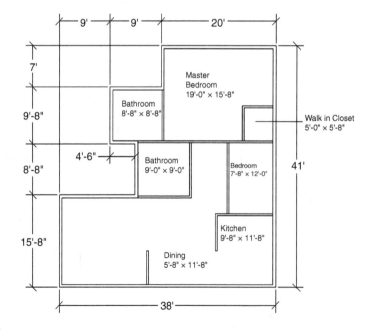

11. Create a 36″ door opening as follows (you can always take a 4″ clearance from the wall). The main entrance, master bedroom, and walk in closet door are at the middle of the wall:

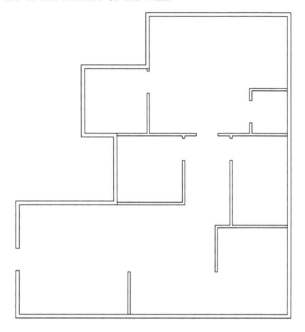

12. Make layer 0 (zero) current.
13. Create the following door blocks using these names (the base point is the lower-left point of the jamb):

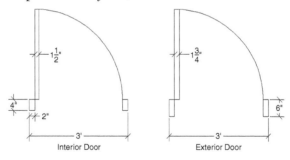

Interior Door   Exterior Door

14. Create the following door block using this name (the base point is the lower-left point of the jamb):

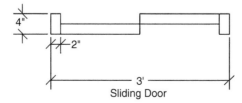

Sliding Door

15. Create the following window blocks using these names (the base point is the lower-left point of the jamb):

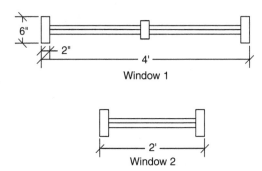

16. Insert the doors and windows in their respective layers to get the following result:

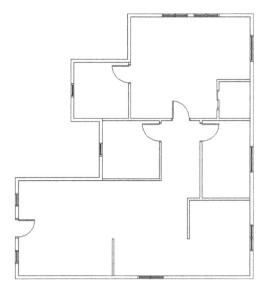

17. Make layer Furniture current.
18. Using Design Center insert the following blocks from Home-Space Planner. dwg, House Designer.dwg, and Kitchen.dwg:

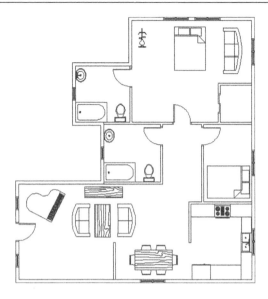

19. Make layer Hatch current.
20. Using Solid hatching hatch both the outside and the inside wall.
21. Using ANSI37 and scale = 100, hatch the kitchen (hint: draw a line to separate the kitchen from the adjacent room).
22. Using the User-defined hatch (switch the Double checkbox on) and scale = 20, hatch both bathrooms.
23. You should have something like the following:

24. Make layer Text current.
25. Create a new text style with the following specs:
     a. Name = Room Titles
     b. Font = Arial
     c. Height = 10"
26. Freeze layer Hatch.
27. Add text using multiline text to add the room titles just like the following, making sure Justify is Middle Center:

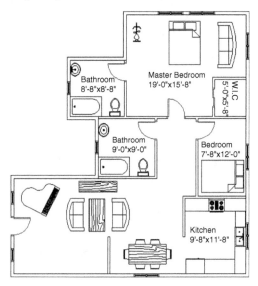

28. Thaw layer Hatch.
29. Select the hatch of one of the two bathrooms. When the context tab appears, locate the **Boundaries** panel and click the **Select** button, then choose the text, press [Enter], and then press [Esc]. Do the same for the other bathroom and the kitchen.
30. Create a new dimension style with the following specs:
     a. Name = Outside Walls
     b. Extend beyond dim line = 12"
     c. Offset from origin = 12"
     d. Arrowheads = Oblique
     e. Arrow size = 12"
     f. Text style = Standard
     g. Text height = 12"
     h. Text Alignment = ISO standard
     i. Primary units = Architectural
     j. Precision = 0' – 0 ½"

31. Make layer Dimension current.
32. Insert the dimensions as shown below (use the Continue command whenever possible):

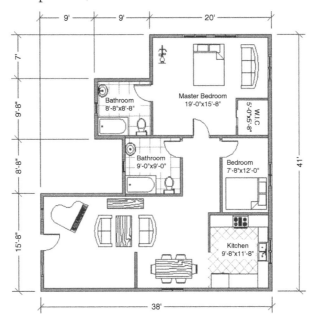

33. Go to Layout1 and rename it to Full Plan.
34. Using the Page Setup Manager modify the existing Page Setup to be as follows:
    a. Printer = DWF6 ePlot.pc3
    b. Paper = ANSI B (17x11 inch)
    c. Drawing orientation = Landscape
35. Erase the existing viewport.
36. Make layer Title Block current.
37. Insert the file ANSI B Landscape Title Block.dwg in the layout using 0,0,0 as the insertion point.
38. Create a copy of the layout and name it Details.
39. Erase Layout 2.
40. Go to the layout named Full Plan.
41. Make layer Viewport current.
42. Insert a single viewport to fill the space, and set the viewport scale to be 3/16″ = 1′, then lock the viewport.
43. Go to layout Details.
44. Create a single viewport to occupy half of the space of the paper.
45. Set the scale to be ½″ = 1″, and lock the viewport.

46. Pan to the entrance making sure to show the dimension, the two windows, and the door.
47. Make the following changes to the Annotative dimension style:
    a. Arrowheads = Oblique
    b. Text alignment = ISO standards
    c. Primary units = Architectural
    d. Precision = 0' – 0 ½"
48. Make the Annotative dimension style current.
49. Double-click inside the viewport.
50. Draw scratch lines at the middle of the two windows and door.
51. Make layer Dimension current.
52. Input the dimensions as shown below, then erase the three scratch lines.
53. Freeze layer Furniture in this viewport, and you will get the following:

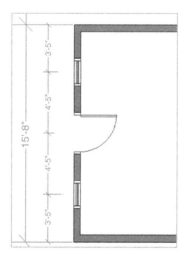

54. Make layer Viewport current.
55. You are still in the Details layout. Create another viewport to occupy half of the remaining area of the paper.
56. Set the scale to ¼" = 1', then lock the viewport.
57. Double-click inside the viewport, and pan to the two windows of the master bedroom.
58. Make layer Dimension current.
59. Draw scratch lines at the middle of the two windows.
60. Make sure you are using the Annotative dimension style and insert the dimensions as shown below.
61. Freeze layer A-Door and Furniture in this viewport.
62. Erase the two scratch lines, and you will get:

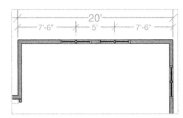

63.   Save and close the file.

10.3 ARCHITECTURAL PROJECT (METRIC)

▪   Take the following steps:
   1.   Start a new project using **acad-Named Plot Styles.dwt**.
   2.   Switch off the grid.
   3.   Set the units to the following:
      a.   Decimal
      b.   Precision 0
      c.   Units to scale inserted contents = Millimeters
   4.   Set up the drawing limits to be:
      a.   Lower-left corner = 0,0
      b.   Upper-right corner = 15000,15000
   5.   Double-click the mouse wheel in order to see the new limits.
   6.   Create the following layers:

Layer Name	Layer Color
A-Door	Blue
A-Wall	White
A-Window	White
Dimension	Red
Furniture	White
Staircase	Magenta
Text	Green
Title Block	White
Viewport	9
Hatch	8

   7.   Save your file in the Chapter 10\Metric folder and name it Ground Floor.dwg.

8. Make layer A-Wall current.
9. Draw the following architectural plan and the partitions inside using these guidelines:
   a. Draw the outer shape using a polyline.
   b. Offset it to the inside using 150 as distance.
   c. Explode the two polylines.
   d. Use the outer wall to draw the inner walls using all the commands you learned in this book. The inner wall is 100.
10. This is the architectural plan:

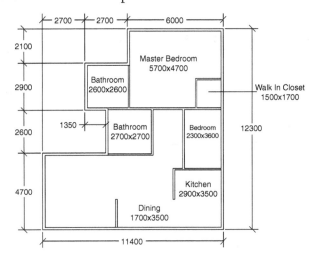

11. Create a 900 door opening as follows (you can always take 100 clearance from the wall). The main entrance, master bedroom, and walk in closet door are at the middle of the wall:

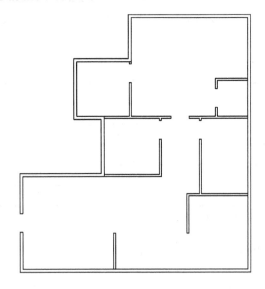

12. Make layer 0 (zero) current.
13. Create the following door blocks using these names (the base point is the lower-left point of the jamb):

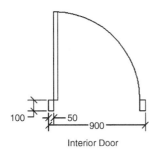 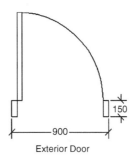

Interior Door   Exterior Door

14. Create the following door block using this name (the base point is the lower-left point of the jamb):

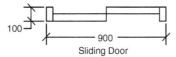

Sliding Door

15. Create the following window blocks using these names (the base point is the lower-left point of the jamb):

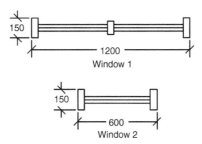

Window 1

Window 2

16. Insert the doors and windows in their respective layers to get the following result:

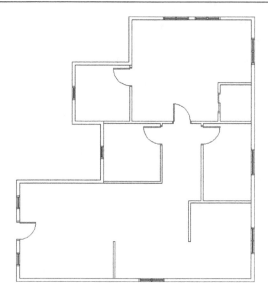

17. Make layer Furniture current.
18. Using Design Center insert the following blocks from Home - Space Planner.dwg, House Designer.dwg, and Kitchen.dwg:

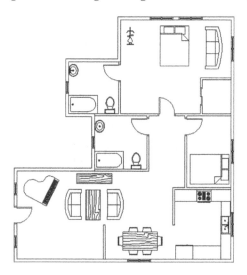

19. Make layer Hatch current.
20. Using Solid hatching hatch both the outside and the inside wall.
21. Using ANSI37 and scale = 2000, hatch the kitchen (hint: draw a line to separate the kitchen from the adjacent room).

22. Using the User-defined hatch (switch the Double checkbox on) and scale = 500, hatch both bathrooms.

23. You should have something like the following:

24. Make layer Text current.

25. Create a new text style with the following specs:
    a. Name = Room Titles
    b. Font = Arial
    c. Height = 250

26. Freeze layer Hatch.

27. Add text using multiline text to add the room titles just like the following, making sure Justify is Middle Center:

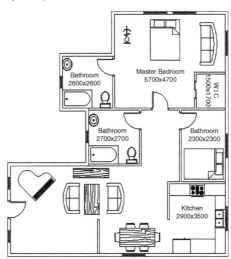

28. Thaw layer Hatch.
29. Select the hatch of one of the two bathrooms. When the context tab appears, locate the **Boundaries** panel, click the **Select** button, and then choose the text, press [Enter], and then press [Esc]. Do the same for other the bathroom and the kitchen.
30. Create a new dimension style with the following specs:
    a. Name = Outside Walls
    b. Extend beyond dim line = 300
    c. Offset from origin = 300
    d. Arrowheads = Oblique
    e. Arrow size = 300
    f. Text style = Standard
    g. Text height = 300
    h. Text Alignment = ISO standard
    i. Primary units = Decimal
    j. Precision = 0
31. Make layer Dimension current.
32. Insert the dimensions as shown below (use the Continue command whenever possible):

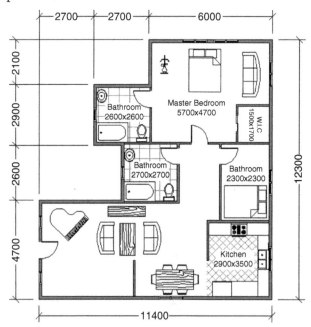

33. Go to Layout1, and rename it to Full Plan.
34. Using the Page Setup Manager modify the existing Page Setup as follows:
    a. Printer = DWF6 ePlot.pc3
    b. Paper = ISO A3 (420x297 MM)
    c. Drawing orientation = Landscape

35. Erase the existing viewport.
36. Make layer Title Block current.
37. Insert the file ISO A3 Landscape Title Block.dwg in the layout, using 0,0,0 as the insertion point.
38. Create a copy of the layout, and name it Details.
39. Erase Layout 2.
40. Go to layout Full Plan.
41. Make layer Viewport current.
42. Insert a single viewport to fill the space, and set the viewport scale to be 1:100, then lock the viewport.
43. Go to layout Details.
44. Create a single viewport to occupy half of the space of the paper.
45. Set the scale to be 1:20, and lock the viewport.
46. Pan to the entrance making sure to show the dimension, the two windows, and the door.
47. Make the following changes to the Annotative dimension style:
    a. Arrowheads = Oblique
    b. Arrow size = 4
    c. Text height = 4
    d. Text alignment = ISO standards
    e. Primary units = Decimal
    f. Precision = 0
48. Make the Annotative dimension style current.
49. Double-click inside the viewport.
50. Draw scratch lines at the middle of the two windows and door.
51. Make layer Dimension current.
52. Input the dimensions as shown below, then erase the three scratch lines.
53. Freeze layer Furniture in this viewport, and you will get the following:

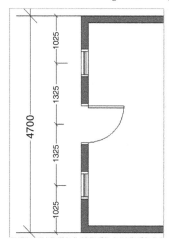

54. Make layer Viewport current.
55. You are still in the Details layout. Create another viewport to occupy half of the remaining area of the paper.
56. Set the scale to 1:40, then lock the viewport.
57. Double-click inside the viewport, and pan to the two windows of the master bedroom.
58. Make layer Dimension current.
59. Draw scratch lines at the middle of the two windows.
60. Make sure you are using the Annotative dimension style and insert the dimensions as shown below.
61. Freeze layer Furniture in this viewport.
62. Erase the two scratch lines, and you will get:

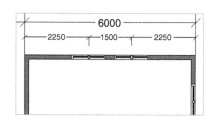

63. Save and close the file.

10.4 MECHANICAL PROJECT – I (METRIC)

■ Take the following steps:
1. Start a new project using **acad-Named Plot Styles.dwt**.
2. Switch off the grid.
3. Set the units to the following:
   a. Decimal
   b. Precision 0
   c. Units to scale inserted contents = Millimeters
4. Set up the drawing limits to be:
   a. Lower-left corner = 0,0
   b. Upper-right corner = 350, 250
5. Double-click the mouse wheel in order to see the new limits.
6. Create the following layers:

Layer Name	Layer Color	Layer Linetype
Centerline	Green	Centerx2
Dimension	Red	Continuous
Hatch	8	Continuous
Hidden	Cyan	Hiddenx2
Part	White	Continuous
Text	Magenta	Continuous
Title Block	White	Continuous
Viewport	9	Continuous

7.   Save your file in Chapter 10\Metric folder and name it Mechanical–1.dwg.

8.   Make layer Part current.

9.   Draw the following plan, section, and elevation of the mechanical part using the following guidelines:

    a.   All lines of the shape in layer Part.

    b.   All centerlines in layer Centerline.

    c.   All hidden lines in layer Hidden.

    d.   Change the **Linetype scale** using the **Properties** palette for both centerlines and hidden lines to 5, except for the two holes at the right and left, which should be 2.

10.   Draw the shape without dimensioning for now:

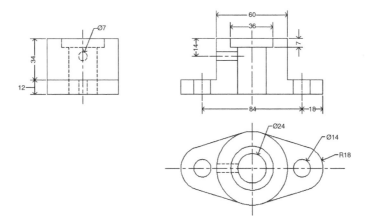

11.   Make layer Hatch current.

12.   Using ANSI31 with scale = 20, hatch the shape as shown below:

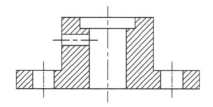

13. Create a new dimension style with the following specs:
    a. Name = Part Dim
    b. Extend beyond dime lines = 3.6
    c. Offset from origin = 1.25
    d. Arrowheads = Right angle
    e. Arrowhead size = 3.6
    f. Text height = 3.6
    g. Text alignment = ISO standard
    h. Primary units = Decimal
    i. Precision = 0
14. Make layer Dimension current.
15. Insert the dimensions as shown below:

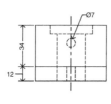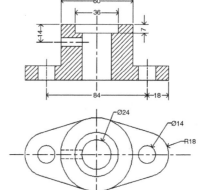

16. Go to Layout1, and rename it Plan.
17. Using the Page Setup Manager modify the existing Page Setup as follows:
    a. Printer = DWF6 ePlot.pc3
    b. Paper = ISO A3 (420x297 MM)
    c. Drawing orientation = Landscape
    d. Make sure Scale = 1:1
18. Erase the existing viewport.
19. Make layer Title Block current.

20. Insert the file ISO A3 Landscape Title Block.dwg in the layout, using 0,0,0 as the insertion point.
21. Make two copies of the Plan layout and name them Section and Elevation.
22. Delete Layout 2.
23. Make layer Viewport current.
24. Insert a single viewport with the following settings:
    a. Set the scale to 2:1.
    b. Lock the viewport.
    c. Pan to the shape as shown below.
    d. You will notice that the hidden lines and centerlines look like continuous lines. To solve this problem, at the Command window type the **psltscale** command, and set this variable to 0. Then type the **regenall** command to regenerate all viewports, and you will get the following result:

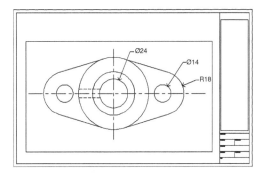

25. Repeat the same procedure to create a section viewport in the Section layout and elevation viewport in the Elevation layout.

10.5 MECHANICAL PROJECT – I (IMPERIAL)

- Take the following steps:
    1. Start a new project using **acad-Named Plot Styles.dwt**.
    2. Switch off the grid.
    3. Set the units to the following:
       a. Fractional
       b. Precision 0 – 1/16
       c. Units to scale inserted contents = inches
    4. Set up the drawing limits to be:
       a. Lower-left corner = 0,0
       b. Upper-right corner = 18″,9″
    5. Double-click the mouse wheel in order to see the new limits.

6. Create the following layers:

Layer Name	Layer Color	Layer Linetype
Centerline	Green	Center2
Dimension	Red	Continuous
Hatch	8	Continuous
Hidden	Cyan	Hidden2
Part	White	Continuous
Text	Magenta	Continuous
Title Block	White	Continuous
Viewport	9	Continuous

7. Save your file in the Chapter 10\Imperial folder and name it Mechanical–1. dwg.
8. Make layer Part current.
9. Draw the following plan, section, and elevation of the mechanical part using the following guidelines:
    a. All lines of the shape in layer Part.
    b. All centerlines in layer Centerline.
    c. All hidden lines in layer Hidden.
    d. Change the **Linetype scale** using the **Properties** palette to be 0.5 for any line you like.
10. Draw the shape without dimensions for now:

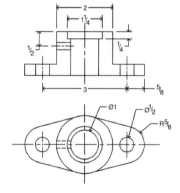

11. Make layer Hatch current.
12. Using ANSI31 with scale = 1, hatch the shape as shown below:

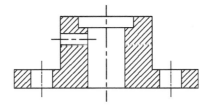

13. Create a new dimension style with the following specs:
    a. Name = Part Dim
    b. Extend beyond dime lines = 1/8
    c. Offset from origin = 1/8
    d. Arrowheads = Right angle
    e. Arrowhead size = 1/8
    f. Text height = 1/8
    g. Text alignment = ISO standard
    h. Primary units = Fractional
    i. Precision = 0 1/16
    j. Fractional format = Diagonal
14. Make layer Dimension current.
15. Insert the dimensions as shown below:

 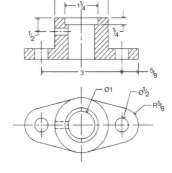

16. Go to Layout1, and rename it Details.
17. Using the Page Setup Manager modify the existing Page Setup to be as follows:
    a. Printer = DWF6 ePlot.pc3
    b. Paper = ANSI B (17x11 in)
    c. Drawing orientation = Landscape
    d. Make sure Scale = 1:1
18. Erase the existing viewport.
19. Make layer Title Block current.

20. Insert the file ANSI B Landscape Title Block.dwg in the layout, using 0,0,0 as the insertion point.
21. Delete Layout 2.
22. Make layer Viewports current.
23. Insert three single viewports and do the following:
     a. Set the scale for the three viewports to 1' = 1'.
     b. Lock the viewports.
     c. Pan to the shape as shown below.
     d. If you notice that hidden lines and centerlines look like continuous lines, at the Command window type the **psltscale** command, and set this variable to 0. Then type the **regenall** command to regenerate all viewports, and you will get the following result:

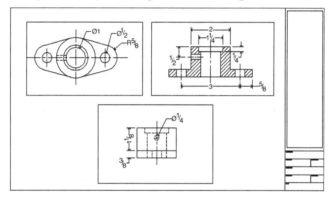

24. Save and close the file.

10.6 MECHANICAL PROJECT – II (METRIC)

■ Using the same methodology we used in Mechanical Project – I (Metric) draw the following project:

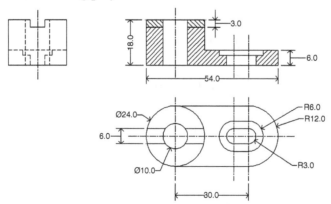

10.7 MECHANICAL PROJECT – II (IMPERIAL)

■ Using the same methodology we used in Mechanical Project – I (Imperial) draw the following project:

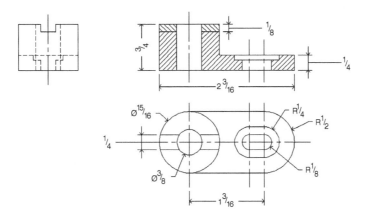

NOTES:

A Appendix

HOW TO CREATE A TEMPLATE FILE

In This Chapter

◇ What are template files?
◇ What are the elements in a template file?
◇ How to create a template file

A-1 WHAT IS A TEMPLATE FILE AND HOW DO YOU CREATE ONE?

- Any company using AutoCAD should look to standardize their work and shorten their production time. Templates can help meet these goals. Standardization includes using the same layer naming, colors, linetype, and lineweight as well as standard text and tables, dimensions and leaders, and standard layouts. Needless to say, companies will use always the same shapes for blocks.
- A good template should include the following:
  - Drawing Units.
  - Drawing Limits.
  - Grid and Snap settings.
  - Layers.
  - Linetypes.
  - Text Styles.
  - Table Styles.
  - Dimension Styles.
  - Multileader Styles.
  - Layouts (including Border Blocks and Viewports).
  - Page Setups.
  - Plot Style Tables.

- Take the following steps to create a template file:
  - Create a new file using the template file *acad.dwt* or *acadiso.dwt*.
  - Gather the information for the settings listed above.
  - Inside the template file, change these settings.
  - From the application menu, choose **Save As/AutoCAD Drawing Template**:

- You will see the following dialog box:

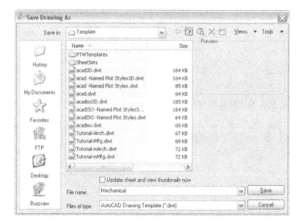

- Input the name of the template file. By default AutoCAD will direct you to the same folder AutoCAD saves its default template files in. You can save your template there, or you can create your own folder, which we highly recommend. To create your own folder, you'll need to specify it for AutoCAD. You can do this using the Options dialog box; see the following:

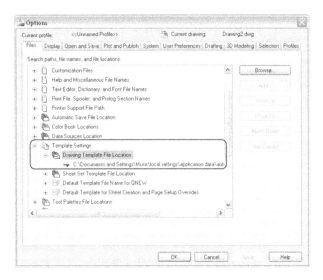

A-2 EDITING A TEMPLATE FILE

- To edit an existing template do the following:
  - Using the application menu select **Open/Drawing:**

- You will see the following dialog box. Using **Files of type** choose **Drawing Template (*.dwt)**:

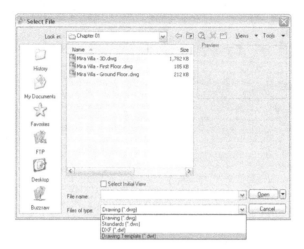

- It will take you to the default Template folder. Select and open the desired template, then make your edits.
- Save it using the same name, or use a new name.

INDEX

www.ingramcontent.com/pod-product-compliance
Lightning Source LLC
LaVergne TN
LVHW062302060326
832902LV00013B/2012